Fix Bayonets!

To my wife Elizabeth, our daughter Charlotte, and granddaughters Harriett and Olivia, who have all had to listen to me thinking aloud and somehow remain interested.

Fix Bayonets!

John Norris

Pen & Sword
MILITARY

First published in Great Britain in 2016 by
Pen & Sword Military
an imprint of
Pen & Sword Books Ltd
47 Church Street
Barnsley
South Yorkshire
S70 2AS

ISBN 978 1 78159 336 3

A CIP catalogue record for this book is available from the British
Library

Typeset in Ehrhardt by
Mac Style Ltd, Bridlington, East Yorkshire
Printed and bound in the UK by CPI Group (UK) Ltd,
Croydon, CRO 4YY

Pen & Sword Books Ltd incorporates the imprints of Pen & Sword
Archaeology, Atlas, Aviation, Battleground, Discovery, Family History,
History, Maritime, Military, Naval, Politics, Railways, Select, Transport,
True Crime, and Fiction, Frontline Books, Leo Cooper, Praetorian
Press, Seaforth Publishing and Wharncliffe.

For a complete list of Pen & Sword titles please contact
PEN & SWORD BOOKS LIMITED
47 Church Street, Barnsley, South Yorkshire, S70 2AS, England
E-mail: enquiries@pen-and-sword.co.uk
Website: www.pen-and-sword.co.uk

Contents

Acknowledgements   vi
Introduction   vii

Chapter 1   A Desperate Act   1

Chapter 2   A New Weapon for the Infantry   4

Chapter 3   The Experience of the English Civil War and Beyond   13

Chapter 4   Time for Change   27

Chapter 5   The Eighteenth Century and the American War of
                  Independence   48

Chapter 6   Back to Europe and Other Conflicts   83

Chapter 7   New Weapons and More Wars   102

Chapter 8   Bayonet Practice   137

Chapter 9   Cut and Thrust   144

Chapter 10   Ceremonies and Other Weapons Fitted with Bayonets   154

Chapter 11   The Twentieth Century and Beyond   170

Where to see Bayonets   204
Bibliography   205
Index   207

Acknowledgements

I would like to extend my gratitude to the many museums which have assisted in compiling this work, especially the various regimental museums such as the Army Medical Services Museum in Surrey, and the Gurkha Museum in Winchester, Hampshire where members of staff have been most helpful. I would also like to say how grateful I am to Peter Norris and Graham Priest, highly knowledgeable collectors of bayonets who provided reference material, and World Wide Arms of Eccleshaw for providing some images. My thanks are also extended to the various re-enactment groups including Tom Greenshields and the Gordon Highlanders 1914–1918 who patiently stood whilst I took photographs, some of which appear in this work; and the Taunton Garrison which depicts the Monmouth Rebellion of 1685. These groups certainly know how to bring history alive. Without such support my task would have been much more difficult. My special gratitude is reserved for John Adams-Graf in America who provided some incredible images for this work.

Introduction

In battle the order to 'fix bayonets' is an ominous command and to those troops who have to carry out the movement to fix their bayonets to their rifles, it means many things. Above all, it says, 'This is it, there is no going back'. It is not an order given lightly and to the infantrymen serving in the trenches of the Western Front during the First World War, the order meant they were going 'over the top' to charge the enemy face to face. It showed the determination to get the job done almost invariably against superior odds. It is a chilling order to receive, especially in modern warfare and yet, incredible as it may sound, it is an order which has been given more times in recent conflicts than one may imagine.

The history of the bayonet is the story of an incredible survivor in terms of weaponry, having been created in the early seventeenth century; more than 300 years later, it is still being issued as an essential part of a soldier's kit during an age when weapons exist which can destroy entire cities. The bayonet has survived it all, having been modified to meet changing designs in weaponry, and is still in use today in its primary role, which is to put paid to any last remaining vestiges of resistance from the enemy. In 1982 during the Falklands War the Scots Guards were given the order to 'Fix bayonets!' Their objective was to clear away the enemy Argentine soldiers holding positions dominating the heights of Mount Tumbledown, and the Guardsmen showed they were up to the task by charging the enemy positions. The bayonet has always had a secondary role as a multi-function tool from opening boxes to prodding for mines, and some modern designs can be used to cut barbed wire.

Modern infantrymen carry anti-tank missiles, hand grenades and machine guns capable of firing hundreds of rounds of ammunition per minute to make them more powerful than ever before. Yet for all that, soldiers still carry a knife-like implement which has a single blade like the men-at-arms who carried swords onto the battlefield in the Middle Ages and earlier. A bayonet is simply a dagger which can be attached to a rifle to use as a thrusting weapon, and in that respect is not that far removed from the Roman legionary who carried a spear or pilum onto the field at the Battle of Metaurus in 207 BC. These same soldiers also carried swords and daggers to stab the enemy, which is exactly the same purpose for which a bayonet is used.

The Romans came to realize that a 'slashing' action with a sword rarely inflicted a fatal wound on an enemy, even one not protected by wearing armour. Such wounds would incapacitate a man but they were rarely immediately fatal. Over the centuries military societies recognized that a stabbing action could puncture arteries, and even a wound 2 inches deep could pierce vital organs and cause internal bleeding and death. This was something the bayonet was ideally suited to do, no matter what the design of the blade, and since it was first introduced military forces have carried them into battle. The armies of Wellington and Napoleon advanced across the battlefields of Europe with their muskets tipped with the bayonet, its use in the Crimean War helped create the popular image of the 'Thin Red Line', and the American Civil War also saw bayonet charges being ordered.

This work sets out to chart the origins of the bayonet from its humble beginnings when it was produced like any knife to the scientifically produced weapon of today, which is ergonomically designed. Over the centuries the bayonet has come to be used for many purposes other than its main function on the battlefield, such as stirring food and chopping firewood. In fact, the author remembers using his issue bayonet as a bottle opener, among other things, during his service in the British army. Troops the world over have traditionally used their bayonets as can openers, and there is an apocryphal story which runs that one day a soldier turns to his sergeant and says: 'Hey! This can opener fits on the end of my rifle.' It may or may not be real, but the fact remains that bayonets have always been used for other purposes and so there is a strong possibility that it could have happened.

The bayonet has never been used to win a battle on its own, but it is the implication for which it stands that has helped turn a battle and led to it gaining an almost mythical status which outranks its use. For all the bayonet charges ever mounted, the number of wounds inflicted compared to other weapons is relatively small in percentage terms. For example, during the American Civil War of 1861 to 1865, an examination of all those wounded and hospitalized revealed that only 922 were recorded as having been admitted suffering from either bayonet or sword wounds. One doctor serving with the Union Army recorded only having ever treated 37 bayonet wounds. This tells only part of the story, because these are the numbers of men sent to hospital for treatment.

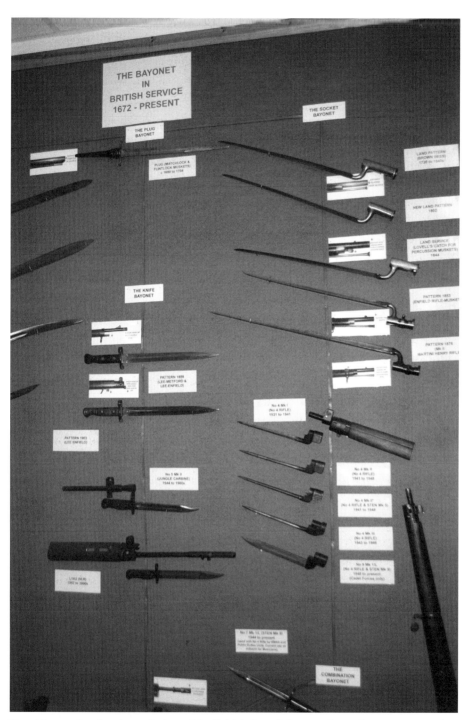

Board layout showing the evolution of the British Army's bayonet from plug type at top, circa 1672, through to modern type fitted to the SA80 rifle in current service.

There would have been many who received bayonet wounds who were treated in field hospitals and not evacuated and properly hospitalized. Those who were killed by the bayonet during a battle would not have been among the figures of these hospital returns and so the true figure would be much higher.

Sir J.W. Fortescue wrote: 'All nations boast of their prowess with the bayonet, but few men really enjoy a hand-to-hand fight with a bayonet. English and French both talk much of the bayonet but in Egypt in 1801 they threw stones at each other when their ammunition was exhausted and one English sergeant was killed by a stone. At Inkerman [5 November 1854] again the British threw stones at the Russians, not without effect; and I am told upon good authority that the Russians and Japanese, both of whom profess to love the bayonet, threw stones at each other rather than close, even in this twentieth-century'. This last incident is obviously a reference to the Russo-Japanese war of 1904–1905, where the two countries fought over territorial gains in China, but does say how troops could be reluctant to "go in" with the bayonet. Indeed, an inspection of casualty records from this war reveals that only 2.5 per cent were inflicted by "swords, bayonets and spears".'

The historian Fred Majdalany wrote extensively about the Second World War and believed there was: 'A lot of loose talk about the bayonet. But relatively few soldiers could truthfully say that they had struck a bayonet into a German. It is the threat of the bayonet and the sight of the point that usually does the work. The man almost invariably surrenders *before* the point is stuck into him.' Another opinion held is that no man was ever bayoneted who had not first surrendered. A British serviceman who fought during the Falklands War in 1982 remembered being confronted by an Argentine soldier with his bayonet fixed. He too had his bayonet fixed to his L1A1 rifle and he came on guard to present the bayonet towards his opponent. The Argentine also came on guard in response. Not wishing to engage in a bayonet fight, the British serviceman shot the Argentine soldier and ended the confrontation. The combat was over in a flash and was one incident in a fight going on all around.

In this modern age of automatic drones, 'smart' bombs and computers it seems incredible that today, even after more than three centuries of existence, the bayonet remains feared and reviled on the battlefield. It is one thing to take cover behind a wall or earth embankment and exchange fire with an enemy, but few troops will stand and face up to a bayonet charge. In 2004 during the war in Iraq, men of the Argyll & Sutherland Highlanders became involved in a firefight with Islamic forces in an action some sources refer to as the 'Battle of Danny Boy'. The men fixed bayonets to their SA80 rifles and charged a mortar position during which they killed around 80 enemy troops. Even more recently Lieutenant James Adamson of the Royal regiment of Scotland was awarded the Military Cross for his part in a bayonet charge in Afghanistan in 2009. Is it by chance that the three most recently recorded bayonet charges should involve Scottish Regiments, and by doing so continue the heritage of the 93rd Highlanders at the Battle of Alma during the Crimean War, where they earned themselves the title of 'the thin red streak tipped with a line of steel.'? The story of the bayonet continues as demonstrated by Corporal Sean Jones of 1st Battalion, The Princess of Wales' Regiment, who was also awarded the Military Cross in 2012 after he mounted a bayonet charge and 'reversed a potentially dire situation' whilst on patrol in Afghanistan in 2011.

The bayonet has also been used as a symbolic image on posters to encourage recruitment into the army, as a symbol of pride and, in the case of Soviet Russia during the Second World War, it was even used as a symbol of resistance against Nazi invasion. Perhaps the most unusual image of the bayonet was when it was used in an advertisement to promote the benefits of Bovril, a hot beef-flavoured beverage in the First World War. There are very few weapons where everybody agrees on the date when it was introduced into military service. Usually there is some degree of ambiguity concerning the

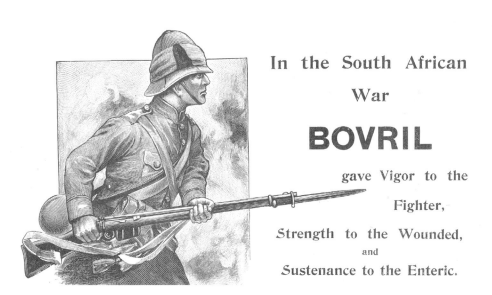

Image of bayonet used for advertising purposes during the Second Boer War of 1899–1902.

precise date, or at least thereabouts, and its place of origin. The bayonet is one of those few exceptions and that is what this book aims to tell.

That is why the reader will appreciate that this is a book looking at the history and development of the bayonet and how it came into military service. There are books on collecting bayonets and identifying the types, which are popular items of military equipment with collectors of such impedimenta. It is the sincere hope that perhaps this book can be used in conjunction with those titles as an aid to collecting bayonets and understanding the use of them in battle. The bayonet has been around for a very long time and there is no reason to believe that it will not remain in service as long as there are infantrymen taking to the battlefield.

Chapter 1

A Desperate Act

Exhausted, half-crazed with thirst, the last of their ammunition fired, the remaining handful of troops capable of standing fixed their bayonets to their rifles. All day the soldiers had been defending their position in a hacienda where they had sought refuge against overwhelming Mexican forces, having stalwartly refused to surrender. Stepping out into the sharp light from the dark interior of the building, the small band of soldiers blinked in the glare before bringing their rifles tipped with their bayonets level and charging into the line of their attackers, who were besieging their little position. It was a last act of defiance born out of desperation and a stubbornness not to surrender. These were men of the French Foreign Legion and they had a reputation for toughness in battle, whatever the odds.

It was 30 April 1863 and the French had been in Mexico since 1861 as part of the Emperor Napoleon III's grand designs to give France an overseas Empire. On that fateful day, the city of Puebla was under siege by French troops commanded by Charles Ferdinand Latrille, Count of Lorencez, an experienced field commander who had seen action in North Africa and the Crimean War. He was short of supplies and sent word for ammunition, food and money in order to continue his action. An escort force for the supply column of 62 Legionnaires was assembled from 3rd Company Foreign Regiment with three officers, including Captain Jean Danjou as the commanding officer. The small detachment set out with the supply column at around 1 am to transport supplies under cover of darkness, which included 3 million francs for the forces at Puebla.

After 15 miles into their journey the order to rest and prepare food was given. Sentries were posted to keep watch for hostile Mexican troops. The time was around 7 am. The water for the troops' coffee had barely had time to boil before the alarm was raised to warn of the approach of a force of around 800 mounted Mexican troops. Captain Danjou knew the tactic to counter a cavalry charge and ordered his men to form a square. The French Legionnaires kept up a steady fire from their Minié rifles, a weapon which had been developed around 1849 by Captain Claude-Étienne Minié and used during the Crimean War ten years earlier, taking a toll on the Mexicans at long range.

With their bayonets with 18-inch blades, capable of piercing a man's abdomen, fitted the Legionnaires were able to prevent their attackers from closing in and managed to repulse several cavalry charges with this tactic. The sun climbed higher in the sky making it unbearably hot and Captain Danjou ordered his men to make their way into a nearby building, the Hacienda Camaròn (a tavern), with adobe walls 10 feet high. The building offered them shelter from the sun and, although they were still in a precarious position, they could at least defend themselves better.

They were able to take cover from Mexican rifle fire behind the walls but the French troops had lost their supplies, were without water and their ammunition was also running low. They realized they could not remain in this position either and it was only a question of time before the inevitable end must come to the battle. The Mexican commander, Colonel Milan, called on them to surrender, during which delivery he reminded them they were surrounded and outnumbered. Danjou responded by declaring: 'We have munitions. We will not surrender.' It was a brave gesture but he supported his defiant mood by declaring he would fight to the death. His men joined him in making a solemn oath, knowing that in so doing they were also condemning themselves to certain death.

By 11 am, about four hours into the fight, the Mexican besiegers were joined by a force of a further 1,200 men, taking their numbers up to around 2,000, pinning down some 65 French. During the fighting the building caught fire but still the French held on stubbornly. The intensity of the Mexican fire increased and Captain Danjou was killed around midday. Command was assumed by Lieutenant Vilain, who continued to inspire his men. Around 4 pm he too fell dead, having been shot when the Mexicans rushed the building. At 5 pm only twelve Legionnaires with Lieutenant Maudet in command were still capable of fighting.

An hour later, with all their ammunition having been fired, Lieutenant Maudet and five men still able to stand fixed bayonets once again, and emerged from the building to make one last defiant but suicidal charge. They were keeping the oath they had made to Captain Danjou, but in the face of such an enemy force it was futile. As they emerged two men, including Lieutenant Maudet, were immediately killed and the remainder were overwhelmed and beaten to death. When the Mexicans entered the building they discovered seventeen wounded and two exhausted but unwounded, whom they took prisoner. The Mexican commander also had to restrain his men from killing these brave unfortunates. One of the French survivors asked Colonel Milan if they might be allowed to depart and take the body of Captain Danjou with them back to their own lines. Recognizing their heroic feat, the Mexican commander exclaimed: 'What can I refuse to such men? No, they are not men, they are devils.' With that he allowed the survivors, some of whom would later die from their wounds, safe passage.

The bayonet charge was a last defiant act by the surviving French soldiers and almost 120 years later in his 1981 book *Introduction to Battlefield Weapons Systems and Technology*, the author R.G. Lee, is echoing what the French Legionnaires at Camaròn must have felt when he wrote: 'The fixing of bayonets is more than a fixing of steel to the rifle since it puts iron into the soul of the soldier doing the fixing.' However, Lee thought that the bayonet was more of 'an emotive rather than a seriously practical weapon'. This opinion has been much debated among many armies over the centuries and, whilst the fixing of bayonets is indeed an emotive gesture, the men who will have to use it are in no doubt that they are about to become engaged in the most serious and personal form of combat, which is fighting hand-to-hand with the bayonet. The men at Camaròn knew they were going to be killed sooner or later and rather than wait, they chose to go out making their own gesture.

Today there are memorials to both sides on the site of the battle. The episode entered the annals of regimental history within the French Foreign Legion and each year the Battle of Camaròn, as it is known in Mexican, or the Bataille de Camerone in French, is marked on 30 April with a parade and Captain Danjou's wooden hand is displayed in honour of those who died with bayonets fixed. The action had been a tactical victory for the Mexicans, but in the end the defence at Camaròn allowed the French to gain a strategic victory at Puebla because the supply column reached Latrille and the city was captured on 17 May.

Chapter 2

A New Weapon for the Infantry

More than 200 years earlier on the other side of the Atlantic in the coastal city of Bayonne in south-west France, located on the confluence of the Nive and Adour rivers near the Spanish border, and in the modern-day Department of Pyrénées-Atlantiques, a weapon design was being created which would have a profound effect on the way wars were fought, as shown at the Battle of Camaròn along with many other wars and battles in between times. The city of Bayonne had long been an important centre of commerce lying on the frontier between France and Spain, and during the Hundred Years War it had been fought over for its strategic position. Its location meant it benefitted from the influence of the Gascon and Basque cultures; and over the centuries that followed the city's fortunes developed, and so too did its strategic importance. In the late seventeenth century the great military engineer and architect Sébastien Le Prestre, Seigneur de Vauban constructed fortifications to defend the city and protect its interests, which were considered to be of vital importance to the French commercial economy and included fishing, trading in spices and whaling. Other local industries were established, including armaments, and the town gained a reputation for the production of quality swords, knives and daggers. These designs were influential and one style of dagger in particular was developed with a long pointed blade and a tapering wooden handle, which allowed it to be inserted into the muzzle of a musket barrel after the weapon had been fired, and thereby turn the weapon into an extemporized spear to stab at infantry and cavalry. This simple, bladed weapon was easy to produce and inexpensive and the style soon became called the 'bayonet' after the name of the city in French or 'baïonette' in Basque.

The story may sound apocryphal, but all reputable sources agree to this being the place of origin where the first designs of bayonet, referred to as 'plug bayonets' – so-called from the design being like putting a stopper or plug into a wooden barrel – were first manufactured. An alternative story surrounding the origin of the term actually takes this expression into account from the French word *bayoner*, which means to 'put a spigot into a cask' and was almost certainly a reference to the tapered shape of the handle. Despite strong evidence which points to supporting the former story, the origin of the bayonet is still subject to open debate. Other centres of weapon production were manufacturing swords and knives at the same time, which leads to the question: why did another centre not produce the design suited to the role? For example, Toledo in central Spain which, by the seventeenth century, had a well-established steel industry and craftsmen had the skills to produce bladed weapons which could be dated back to around 500 BC. Indeed, the Roman army had adopted a Spanish style of sword they called the *gladius*, from the term *Gladius Hispaniensus* or Hispanic Sword, to arm their troops.

Travelling west to the town of Solingen in the North Rhine-Westphalia region in modern Germany, one discovers a tradition of metal working which dates back 2,000 years. The local blacksmiths in the seventeenth century were renowned for the quality of the knives they produced. It was an industry that had to be protected, and in the fifteenth century this was recognized and the town was fortified. Such was the reputation of the craftsmen's knife-making skills that the town is still referred to as the 'City of Blades' and today that tradition is carried on by companies such as Dreiturm, DOVO Solingen, Wusthof, J.A. Henckels and Eickhorn-Solingen, which still produce knives of exceptional quality and also bayonets for the military. By comparison the bladesmiths in England were making swords and knives, but the country's steel-making process was not nearly as well developed as in Europe. Bladesmiths from Solingen travelled across Europe taking their knowledge with them and expanding their markets for the weapons. Some of these specialists are understood to have settled in Shotley Bridge in County Durham in England in the late seventeenth century. At around this time a blade factory was established in Vira in Sweden by Admiral Fleming, to supply bladed weapons to the Swedish military. In Italy there were steel production centres and in the fifteenth and sixteenth centuries they were also making quality swords and knives. Damascus steel, a method of producing steel which is believed to have originated in India around 300 BC, was used to produce swords and knife blades until around 1700 AD. So why was it that the blades of Bayonne were selected for use by the military? Under other conditions and at another time the blades of any one of several other European cities could have been chosen, but as it turned out it was Bayonne that had the good fortune to have produced the right design. The French army at the time was the most powerful in Europe and engaged in wars on its various borders with Spain and Belgium. Armies required new innovative weapons and tactics if they were to remain dominant, and the simple bayonet would give the French army that edge over its opponents.

Compared to other military weapons the history of the bayonet is not very old and because so many sources agree on its year of origin and place of origin, tracing its history is fairly straightforward. Any existing discrepancies can be explained satisfactorily. One of the earliest written references to the bayonet appears in the work *A Dictionarie of the French and English Tongues*, published in 1611, in which the author Randle Cotgrave describes the entry for 'bayonet' as being: 'A kind of small flat pocket dagger, furnished with knives; or a great knife to hang at the girdle, like a dagger.' Whilst not directly alluding to the fact that the bayonet originated in the city of Bayonne or crediting it as the centre of production for the design of the bladed weapon, the entry does provide the earliest term for a specific type of weapon from which the bayonet would evolve. By the mid-seventeenth century more descriptions of and references to the 'bayonette' were beginning to appear, and the bayonet is increasingly mentioned in a number of texts. Among these are the writings of Pierre Borel in 1655. Born around 1620, Borel trained to become a doctor and, as a highly educated man, he interested himself in a wide range of subjects and wrote notes on his observations. One subject that attracted his attention was the long-bladed knife produced in Bayonne. Earlier references to similar weapons are not entirely specific and these descriptions may only be recording daggers or knives of unusual local design, which are like the bayonet in style. For example, Étienne Tabourot, born in France around 1549 into a well-educated family, wrote extensively and in 1582 he was appointed to the office of the King's Attorney. One of the works he prepared was the Tabourot des Accords, sometimes known as the Seigneur des Accords. In the work is mentioned an account of trade from the city of Bayonne and refers to 'bayonettes' as one of the products traded by the city. Beyond that, though, not much is noted and he may be referring to the knives and blades produced and ascribing to them the collective term 'bayonettes'.

The strongest theory as to how and why the bayonet was developed is the suggestion that it may have been inspired by hunters who inserted the handles of their knives into the barrels of their guns to turn them into rudimentary spears. This was at a time before the weapon had been drawn to the attention of the military, and its style was probably better suited to civilian usage where it would have proved handy during hunting trips. Indeed, hunting was a way of life in the Franco-Spanish border region to provide food and if a hunter missed his target, such as a wild boar, or wounded it he could push the handle of his dagger or bayonet into the muzzle of his weapon. This would be useful against wounded animals such as a boar, which have been known to turn and attack its would-be hunters – in which case, the combination of dagger and gun would be ideal to impale the quarry.

The shape of the first bayonets were long flat blades tapering down to a point in a style referred to as 'spear point', a design which was commonly used in sword-making from ancient Egypt through to the Italian short sword known as the 'cinquedea' during the fifteenth and sixteenth centuries. The blades of these first bayonets were individually hand-forged by blacksmiths, originally as single items; but when their significance as military weapons was recognized and as armies increased in size, so many more craftsmen would have been engaged in their production on an industrial scale to increase manufacturing output to meet demand. To this day, many bayonet designs still retain their dagger-like form, which serves as a reminder of its humble origins. There have been variations over the centuries, but the flat-bladed design has been proven to be the best shape to have. A wide range of long, thin-bladed knives and daggers with spear point tips known variously as stiletto, rondel or misericordia had long been used by foot soldiers as stabbing weapons in hand-to-hand fighting on the battlefield. The English, for example, used a type known as the 'ballock' or 'bollock' dagger in the thirteenth century in reference to a pair of bulbous protuberances incorporated in the design of the wooden handle. The weapon would later become known by the more polite term 'kidney' dagger in Britain during the nineteenth century. The Scottish dirk has a similar blade design, but of all the styles produced, it was the dagger made in Bayonne which would be used to equip first the French army, and later other European armies, including England. The style did not disappear altogether and during the Second World War the Fairbairn-Sykes dagger was developed for use by British commandos, and the lineage that influenced its design can be seen even down to the handle, which is almost the same as the first plug bayonets. If imitation is the sincerest form of flattery, then the Fairbairn-Sykes dagger was paying homage to bladesmith designs of long ago.

The blades of the early bayonets were rough and without any pattern, but they were solidly made and reliable, which soldiers would need, and hunters would have appreciated before them. Some individual blades were adorned with decorations and some even had highly decorative handles inlaid with mother-of-pearl or ivory, and even gold or silver. These would have been specially commissioned items to indicate a symbol of status or rank, but they were nevertheless functional items. The original plug bayonet resembled a miniature spear, having a similar shaped blade and a tapering wooden handle to allow it to fit into any size musket barrel, just like the plug in a wooden cask or barrel. This style of weapon was known to Louis de Gaya who served as a captain in the Charlemagne Regiment of the French Army, and recorded the weapon in his work *Traité des Armes*, published in 1678, mentioning that: 'The bayonet is about the length of a dagger. It has no guard, but only a wooden handle length of eight to nine inches. The blade is pointed and cutting, one foot long and a good inch wide. The bayonet is of good utility with the dragoons, the fusiliers and with the soldiers who are often ordered to go to war; because when they have made their discharges, and have no ammunition, they put the handle in the muzzle of their fusils and defend as well as with a partisan.' From this description we see how the bayonet is already being seen as an important weapon for close-quarter fighting and self-defence. The account also tells us how the size of the bayonet could vary in length and some blades could be exceptionally broad.

The shape of the blade could vary in style from the standard tapering spear-like point, and designs with some blades featuring a slight curve near the tip to provide a double edge were experimented with. Other styles would be developed and for their size, some up to 18 inches in length, the early plug bayonets could be remarkably lightweight. In fact, many styles weighed less than 1lb (0.5kg), but they still retained a good solid feel to them as a weapon intended for use by soldiers should be, if it is to last in battle. Gaya's account also confirms the observation of other writers who describe how, after having fired their muskets, the troops pushed the handles of their bayonets into the muzzle end of their muskets. He also recounts how the regiments of dragoons (which were essentially mounted infantry) used the musket fitted with a bayonet whilst fighting on foot in the same method as though it were a partisan, which was a form of short pike or spear.

The blade of the plug bayonet could exceed a length of 12 inches and some troops were of the habit of sharpening the edge in some designs to make them a useful all-round tool for chopping and cutting as well as stabbing. The broadest part of the blade closest to the handle was called the 'ricasso' and fitted with a crosspiece or cross-guard known as a 'quillon', which was held in place by a metal ring called a 'ferrule'. The tapering wooden handle varied in length, but an average was usually around 12 inches, also. The early styles did not have a cross-guard or quillon and some types issued to the 'rank and file' troops of the infantry had a handle with a bulbous bulge known as a 'swell', which prevented it from being pushed too far into the barrel where it could become stuck. The tapering style was meant to prevent this by only allowing the handle to be pushed into the barrel up to a point which would accommodate it. This was a universal 'one-size-fits-all' design, but some bayonets were fitted with handles that incorporated metal collars or rings to prevent them from being pushed in too far. In the event that the handle did become jammed in the barrel, after being used to stab a man or horse – the force of which could push it further back into the barrel – the musketeer could knock the bayonet out by tapping on the quillons. The handle fitted over the tang, which was the upper portion of the blade, and formed into a point. The handle was held in place by a metal pommel, which was usually plain brass, but some designs with fancy decorations were produced as individual items, probably to denote status.

Exactly who first thought of the idea of pushing the handle of a dagger into the muzzle of the gun barrel has long been lost in the mists of time. It could be that it happened purely by chance, and it has been opined that it may have first been used in Spain. One of those who agreed with this idea was the nineteenth century weapon historian James Robinson Planché, who believed it may have happened as early as 1580. This may well have been the case and, given the proximity of Bayonne to the Basque region on the border with Spain, the practice could easily have spread among hunters who recognized the usefulness of such a combination.

Among those to disagree with this theory was Rodolphe Schmidt, another weapons historian, who believed the practice did not evolve until perhaps fifty years later. This would place the practice as having developed around 1630, but written references to daggers already being used in this way with muskets dating from an earlier period casts an element of doubt on this notion. The practice had to start somewhere but, like the originator of the idea, it is simply not known. The person with the strongest case to be credited as being the instigating force behind the introduction of bayonets into general service throughout the French army is Lieutenant Colonel Jean Martinet, who held the position of Inspector General of the army and would have been influential in suggesting its formal introduction sometime in the 1660s. Martinet was never a popular figure and the troops despised him for his harsh training regime. Regardless of this notoriety, he did transform the French army into a more efficient fighting force and one of his ideas could well have been the introduction of the bayonet. The date of 1660 is only thirteen years after it was devised and, as the French army was being reorganized and receiving better training, it would have made perfect sense to an organizer like Martinet, who would have recognized the benefits of the weapon and been able to use his position as an opportunity to introduce the bayonet. His date of birth is not know precisely but he is understood to have been killed, probably intentionally by a French soldier, during an attack at the Siege of Duisberg in 1672. Another legacy he left was the theory of volley fire to maximize the effect of musketeers, and this would be copied by other armies and developed even further.

All weapon historians agree that Bayonne is the place of origin for the bayonet and that it was first introduced into military service in 1647 and in view of such overwhelming support, who can argue? It is quite possible that Planché may have reached his conclusion concerning his date by basing his findings on remarks referring to the types of knives made in Bayonne and the reputation of quality. The author Harold L. Peterson in his work *The Book of the Gun* certainly holds this to be the case. The first conclusive references to 'bayonettes' being used by the military come through in writings from around the 1640s, again something agreed on by Peterson, and bayonets are recorded as being used by French troops serving with the Army of Flanders during the closing stages of the Thirty Years' War (1618–1648), which makes this army the first to use the bayonet in a military capacity on campaign. For the time being, at least, the French now had the upper hand by being the first and only military force to know about the bayonet and appreciate its potential. The first plug-type bayonet when fitted may have turned their muskets into replacement pikes but the action also prevented the weapons from being fired, which does suggest that at this point in its history the bayonet would have better served as an accoutrement for parades. Fortunately, someone had the vision to see beyond this drawback and recognize its full potential on the battlefield.

Credence is given to the year of 1647 as the date when the bayonet entered service with the French Army by writings of Chevalier Jacques de Chastenet, Seigneur de Puységur, who at the time was serving as a Colonel of the Régiment de Piedmont, in which he recalled an incident during an engagement in Flanders during the Thirty Years' War in modern-day

Belgium. He wrote how: 'For me, when I was in command at Bergue in Ypre, Dixmund and Laquenoc, all the parties that I sent out passed the canals in this fashion. It is true that soldiers did not carry swords, but had bayonets with handles one foot long, and the blades of the bayonets were as long as the handles, the ends of which were adapted for putting in the barrels of the fusils [muskets] to defend themselves, when attacked after they had fired.' What is interesting from these records is that he is not implying that the bayonet is a new weapon but rather, he mentions how it is now being carried and used by troops for the purpose of self-defence. These observations imply that the bayonet was something which may have once been in limited use for some time but was now gaining wider acceptance and beginning to be issued as standard equipment, and would eventually come to be carried by all troops armed with muskets.

Lieutenant Colonel Belhomme in his observations on the use of the bayonet believes that it may have already been in use with some French troops in 1642. One reference source suggests that it may have been around 1640 that a group of musketeers, finding themselves without the support of pikemen, jammed the handles of their daggers into the muzzles of their muskets. From this incident the idea caught on and then spread to other troops. The weapon historian William Reid maintained that 'From the second quarter of the seventeenth century at the latest, it [the term bayonet] has been used almost universally for knives and swords utilized to make a firearm into a feasible thrusting weapon'. The fitting of plug bayonets into the muzzles of muskets, whilst providing the musketeers with the means to protect themselves quickly against attack by cavalry, had one serious drawback. With the bayonet jammed in the muzzle the muskets could not be reloaded until they were removed. Cavalry soon came to realize that when bayonets were fitted the muskets were not loaded and could not be loaded and fired until the bayonet was removed. Until such time they knew they were safe from being shot at.

Cavalry commanders such as Major-General Thomas Morgan, who whilst campaigning in Scotland in 1654, ordered that his troopers should throw their carbines or pistols at the enemy after they had fired and then 'fall on with the sword.' An act of disdain, perhaps, but certainly very wasteful with firearms which were costly to replace and capable of being used many times more if they had been returned to their holsters. At the Battle of Dettingen in 1743 the cavalry was still practising the wasteful tactic of throwing pistols at the enemy after the riders had fired their weapons, as a British eyewitness later wrote: 'They [The French] rode up to us with a pistol in each hand, and their broad swords slung on their wrists. As soon as they had fired their pistols they flung them at our heads, clapped spurs and rode upon us sword in hand. The fury of their onset we could not withstand so they broke our ranks and got through; but our men immediately closed and turned about, and with the assistance of a regiment… who were in our rear, the French horse being between both, we killed them in heaps.' What is interesting in these two examples is that in each case the cavalry does not appear reluctant to move in close to engage the musketeers. In the first instance the musketeers are not yet equipped with bayonets, and so the riders have nothing to fear and are able to get in among the infantry with their swords. Almost ninety years later the musketeers now have bayonets to protect themselves against cavalry attack and yet they are still attacked with swords. It would seem that the infantry had yet to develop proper tactics to repel cavalry, which in turn had yet to learn lessons when attacking musketeers equipped with bayonets. When they did, each would have to change their tactics.

Examples of plug bayonets are held in private collections and museums where the transition of the bayonet can be examined. Modern historical re-enactors depicting the period of the seventeenth century show how plug bayonets were fitted and used and this interpretation is based on well-researched fact through documents of the time. In the 1947 light-hearted comedy film *The Ghosts of Berkeley Square*, the story is set in the period of 'Marlborough's Wars' and a scene shows soldiers with muskets fitted with plug bayonets to illustrate how they were used in ceremonial duties. The scene provides a good representation and shows how any European army in the early eighteenth century might have looked.

Chapter 3

The Experience of the English Civil War and Beyond

In England civil war broke out in 1642 and turned into a conflict which would tear the country apart as fighting spread and regions chose to support either the Royalist forces of King Charles or the Parliamentarian forces. The first engagements between the two sides in 1642, such as Edgehill on 23 October, proved inconclusive and the year ended with neither side dominant. Oliver Cromwell, who would emerge as the leading commander of Parliamentarian forces during the Civil War, recognized a number of weaknesses in their forces and suggested creating a properly trained army. During the next two years as the war continued numbers of Parliamentarian troops, which would become known as the 'New Model Army', were being trained and equipped. Finally, on 14 June 1645 at the Battle of Naseby, Cromwell and his New Model Army showed what they had learned and gained a victory over the king's army. This was the first true standing army in England and comprised the four main elements of any army at the time: artillery, musketeers, pikemen and cavalry, which Cromwell believed in and held the personal conviction that a man on a horse was very powerful. Some Royalist commanders, such as Prince Rupert who commanded cavalry for King Charles I, and some among the rank and file troops had served in the wars in Europe. Although these men were experienced in battle and familiar with tactics and weapons of the day, it is almost certain that for the most part they did not know about the bayonet, which was still not yet in wide use on the Continent. The war virtually isolated the country from Europe in terms of exchange of military ideas and this meant the bayonet remained unknown to the armies of both sides in England, except perhaps in a few isolated cases where men may have been familiar with the weapon. Mainland Western Europe was militarily active on a wide scale with a number of countries including France, Sweden, Austria and Denmark, engaged in fighting the conflict which would become known as the Thirty Years' War and these troops, especially the French, were accepting the bayonet as a new weapon for musketeers.

The English Civil War continued into mid-1646 when a truce was concluded, but it was only a temporary peace and fighting broke out again in 1648. In that year on the European Continent, the Thirty Years' War was concluded and the belligerent nations used the ensuing peace to put into practice some of the many lessons they had learned during the war. With England once more at war with itself in a second civil war, troops on both sides found themselves using the same weapons and same tactics, but still no bayonets. The New Model Army was the only real innovative creation in the country at that time. On mainland Europe new regiments with specialist roles such as grenadiers and fusiliers were being formed and these would come to be trained in the use of new weapons such as the flintlock musket, hand grenade and, of course, the bayonet. By now the bayonet had already been used on military campaign in France for the first time and whilst its use may only have been limited, the fact remains it was being used and that by itself was to prove a turning point from which new tactics would emerge. In 1651 the English Civil War ended with the Parliamentarian army and Cromwell's New Model Army emerging victorious. King Charles I had been executed in 1649 and his son Prince Charles had fled into exile in Europe. The British Isles was now a military state declared a Commonwealth, with Oliver Cromwell as its Lord Protector. Oliver Cromwell died in 1658 and Prince Charles returned to become King Charles II in 1660, bringing with him his military commanders who had served him in exile and between them they would raise new regiments and introduce new weapons, one of which would be the bayonet.

When Charles returned to England he came back to a much-changed land and one in chaos. With him restored to the throne as king, order would return but this would take time. Changes also had to be made with the military and with a standing army now in place to protect the country, the troops would need new weapons to replace the equipment , which was almost worn out, and also the latest type of weapons. In 1662 an order was placed for a stock of 500 bayonets which, until that time, were still virtually unknown in England except for perhaps a few troops who may have used them during their service as mercenaries in the wars in Europe. This order may have been the first time bayonets had been purchased for use by troops in England and indeed, the idea to buy bayonets may have stemmed from either King Charles II himself ,or someone in his retinue who could have seen bayonets for themselves during their time in Flanders. If this was the case, then it would certainly make perfect sense to introduce the weapon, which would allow the English army to keep up with developments in European armies, some of which, such as the French army, at that time had already been using the bayonet for some fifteen years or so.

Delivery of the first order of bayonets for Charles' army was made in March 1662, but all was not right with them and their condition was most unsatisfactory. On 14 March it was recorded how: 'Ordered that the French pikes and ye short swords or Byonettes that lately were recd. From Dunkirk be surveyed & an accompt presented to the office of their defects to ye end a Contract may bee made for the speedy repaire.' From the date

Recreated infantryman from Monmouth Rebellion with plug bayonet fitted to musket.

of the transaction it would be safe to assume that the bayonets were of the plug type, but reading the document tells us that the condition of the weapons were far from pristine. It may be that funds for purchasing new equipment were limited or that someone was trying to make money out of the deal by purchasing bayonets which had been used, and pocketing the money left over from having bought them cheap. It has even been suggested that the French King Louis XIV may have made a gift of them but, given their condition, it was far from being a generous gesture.

Rather than return the items, however they were obtained, it was decided they should be refurbished and issued to troops. Three days after the assessment it was decided that they should be sent to three specialist London-based cutlers for work to be completed. Joseph Audeley and Samuel Law would work on 200 bayonets and the last tranche of 100 weapons would be sent to Robert Steadman. The letter dated 17 March stated that: 'Ordered yt ye Byonettes lately recd from Dunkirk to be issued to the psons foll. To be by them made clean and repaired & returned with 10 daies space at 14d apc.'

From this document we learn that the work had to be completed in ten days and the price for the work was set. The agreed cost of 14d (one shilling and two pence in pre-decimal money) per item seems quite expensive considering only refurbishing is required and not the manufacture of a whole new item. Depending on the size of the contractor's workshop, this work would have easily been completed in the time allocated. The contractors Audeley and Law would each have earned over £11 and ten shillings (£11.50: approximately £517.50 in modern money) for their work, and Steadman would have earned over £5 and fifteen shillings (£5.75: approximately £258.75 in modern money). The companies engaged in refurbishing the bayonets were experienced in dealing with knives and, whilst the bayonets would have been new items to them, the workers would have known how to tackle the task.

These bayonets being of the plug type with tapering handles were capable of being fitted to the muzzles of the muskets in service with English troops. Unfortunately, the muskets in service were very old and the metal forming the barrel had worn dangerously thin from overuse. When the bayonets were fitted the barrels of some muskets could not take the stress, especially when pressure was exerted during use when a man stabbed at a target with a bayonet; this could often cause the fragile barrel to split. The bayonets were useful but the ageing muskets throughout the army had to be replaced with new weapons, which would lead to the introduction of the flintlock musket throughout the army. It was possible to use the plug bayonet with the matchlock musket, but the flintlock musket was much better suited to the combination. This was also an ideal opportunity to implement other changes and raise new regiments.

Over the next twelve years following the restoration of King Charles II, further changes were made in the English army; and despite the poor beginning, there was a gradual appreciation of the usefulness of the application of the bayonet; and on 2nd April 1672 a Royal Warrant was issued by King Charles II which proclaimed: 'Our Will and Pleasure is, that a Regiment of Dragoons, which we have established and ordered to be raised, in Twelve Troopes of fourscore in each beside officers, who are to be under the command of Oure most deare and most entirely beloved Cousin Prince Rupert, shall be armed out of Our stores remaining within Our office of the Ordinance… the soldiers of the several Troopes aforesaid, are to have and carry each of them one match-locke musket, with a collar of bandoleers, and also to have and carry one bayonet or great knife.' Six years later it is recorded that King Charles II rewarded two Englishmen in 1678 for having invented the 'socket' method of attaching the bayonet, which enabled it to be attached on the outside of the barrel. This is almost certainly a misunderstanding of the term and the new design would probably have been the ring bayonet, which was just beginning to enter use with the French army that same year and was a move which was later copied by other European armies. Socket bayonets, in the true sense would not appear for more than another forty years. It may be possible that the two men King Charles II rewarded were simply copying the trend in Europe and claiming it for themselves as an original design, having heard of it from sources in Europe. The ring bayonet had a number of versions and several claimants purporting to be the inventor, including at least one army officer. We know this to be the case because in 1678 de Puységur recorded in his work *Art de la Guerre* that he had observed troops carrying swords which had no guards, but which were fitted instead with a pair of brass rings attached to the handle which could be passed over the muzzle of the barrel to allow the blade to be attached to the outside of the musket barrel, rather than being jammed into it.

Artillery was a powerful element of any army, whether deployed in open battle or used in sieges against towns, castles or walled cities, and as such it was vitally important in any battle. Unfortunately, its weight and bulk meant it could only be moved at a very slow pace which rendered it highly vulnerable to attack and ambush whilst on the march. This vulnerability had long been recognized and the risks did not diminish when the artillery was deployed on the battlefield where the problem continued once the opening salvoes had been fired by the batteries and battle had commenced proper. On the battlefield artillery usually remained where it had been deployed prior to the outbreak of fighting and almost always stayed in these positions for the duration of the battle. As the main army with its infantry units moved forward, this left the guns unprotected and made them attractive targets subject to cavalry charges. This meant the gun crews had to defend themselves, which was not always possible because they frequently lacked adequate weapons with which to protect their positions. France was the first to recognize that artillery needed a special bodyguard dedicated to protect the guns against attack, but such a move would mean taking troops away from their primary role on the battlefield. Nevertheless, military planners knew that in order to keep the guns safe they had to have units specifically designated to protect them and these were the fusiliers.

One of the first attempts to provide the artillery with some kind of bodyguard was to raise special units known as dragoons. In effect these troops were mounted infantry and as such were armed with musket and sword as infantrymen. Dragoons were not really detailed specifically as artillery bodyguard, but being mounted meant they could ride quickly to any part of the battlefield to give support. In an emergency they could ride to the artillery positions where they would dismount and fight as infantry to repel an attack. Dragoons were used across Europe and regiments had even served during the English Civil War from 1642 onwards, with both Parliamentarian and Royalist forces making use of them. The muskets carried by dragoon troops were of the 'flintlock' type as opposed to the standard matchlock weapons which were discharged by means of applying a burning taper or 'match' to the priming pan. In that respect the matchlock musket was like a small piece of artillery; being loaded in the same manner and fired by a match. The flintlock weapon, however, was fired by means of a flint striking a steel cover to produce a spark to ignite the charge of powder. This dispensed with the need to have a continuously burning match. Flintlock weapons were not only considered to be safer when guarding large stores of gunpowder, but were also more reliable in damp conditions. Flintlock weapons used by dragoons were generally much shorter than matchlock weapons, having a barrel of only 16 inches in length, which made loading more expedient whilst remaining mounted. These shortened weapons had the same calibre as a full musket bore and became termed 'dragons' – a corruption after the troops who used them.

Another type of specialist formation to be raised at this time was the grenadier, companies of which were also armed with the flintlock musket and the bayonet.

Recreated infantryman with plug bayonet fitted ready to receive cavalry, circa 1685.

The historian Saul David believes that along with the flintlock it was the bayonet, which transformed the musketeer from 'a lumbering, vulnerable and ineffective soldier to something resembling the all-purpose infantryman that still exists in the British Army today.' He is, of course, absolutely correct and furthermore agrees with the generally held opinion that it was this weapon which marked the beginning of the end of the role of the pikeman on the battlefield, by providing the musketeers with the means to protect themselves defensively rather than relying on the pike block. Some sources go further by claiming the introduction of the bayonet marked the end of the medieval period and the beginning of modern warfare. It was not just in England that the appearance of the bayonet would transform soldiers; it would also help change armies around the world.

Changes were also happening in France where King Louis XIV in 1671 ordered that a new specialist unit be raised to join the ranks of the army and this was known as the 'Régiment des Fusiliers'. The troops of this new formation were tasked with the specific duties of protecting the artillery and all its impedimenta, including horses and stocks of gunpowder. The troops were armed with short flintlock muskets, known as fusils, which were much more expensive than the matchlocks but more reliable to use. A French Jesuit priest by the name of Father Gabriel Daniel was an official historian and chronicler to the royal court of King Louis XIV and he noted of the bayonet in that year how: 'This weapon is very new to the troops'. He continues his impression in the first person of the King by writing: 'I, The King, first gave this weapon to the Fusileers Regiment, created in 1671, and since called the Royal Artillery Regiment. The souldiers of this regiment carry the bayonet in a small sleeve, as with the sword. I have since given some to other regiments for the same purpose, that is, to put at the end of the musket on occasion.' The bayonet fitted to muskets would have been an added advantage to these troops when carrying out sentry duty to guard artillery and other stores.

The fusiliers, sometimes written as fuzileers, were well trained and, being drilled as infantry, they were reliable troops. At around this time the dragoon regiments across Europe were being absorbed into cavalry units, a trend which also spread to England. In 1673 the first such English unit to be styled as fusiliers was raised at Bois-le-Duc for service in Holland as Sir Walter Vane's regiment. The following year another regiment was raised at Bois-le-Duc and known as either The Irish Regiment or Viscount Clare's Regiment. Both of these regiments would eventually be transferred to service in England

and later evolve to become The Royal Warwick Fusiliers and The Royal Northumberland Fusiliers respectively. The infantry now comprised four types of troop: pikeman, grenadier, fusilier and musketeer. With the introduction of the flintlock musket and the abandoning of the pike, all these types of troops would gradually come to serve in the role of infantry and each man would carry a bayonet for that purpose. The French historian and philosopher Voltaire (1694–1778), at a much later period, wrote of the event and reflected how: 'The use of the bayonet at the end of the musket, is of the king's institution. Before him, one made some use of it, but there were only some companies which fought

The musket and plug bayonet replaced the pike.

with this weapon. It was not widespread, nor put to much use. It was left to each general. The pike passed from being the most formidable weapon. The first regiment which had bayonets, and has used them, were the Fusileers established in 1671'.

With the first regiments now equipped with the new bayonet, it was only a question of time for the rest of the army to be issued with the weapon. Numbers had to be produced and the knife-making centres would have benefitted from the orders to produce bayonets for the army. Supplies would have still been brought in from France, but it was better to have reliable supplies from producers in one's own country should stocks be cut off due to war. The army had traditionally relied on the pike blocks to protect the musketeers using a tactic known as 'push of the pike' when they were reloading and at their most vulnerable from attack. The pike block used their weight of numbers and mass to break up an attack directed against the musketeers, but slowly and surely as the bayonet entered service in greater numbers, the musketeers were then in a position to protect themselves. By 1680 the 2nd

Tangier Regiment (later to become the King's Own Royal Regiment) is recorded as being equipped with 'strap-dagger' bayonets with blades of 12 inches long. These were probably also of the type known as 'ring bayonets', which had already been in service for at least two years with some European armies. The loose-fitting rings attached to the handles allowed them to be slipped over the end of the barrel of the muskets and gave them the appearance of being strapped to the muskets. In 1678 a certain Phillip Russell was paid the sum of eight guineas (£8 and eight shillings) to make 'a new sort of bayonet'. The warrant was signed by the Duke of Monmouth, but unfortunately not many more details than that exist.

There was no formalized bayonet drill in 1685 and infantry used it the best way they could.

Monmouth was the illegitimate son of King Charles II and at the time of his approaching Russell, who may have been a blacksmith, he was fighting in the Franco-Dutch War of 1672–1678. As a highly experienced soldier Monmouth may have been asking Russell to make him some of the new ring bayonets he may have seen in Europe in order to equip his troops in a similar manner. It is known that these bayonets were being used by the French at this time, because an account of the attack against the city of Valenciennes by units of the French army in March 1677 tells us how: 'The cavalry which closed on the place came to the charge and pushed back our people until under the gate [of the city]; but the musketeers, having put their bayonets in their fusils, walked to them, and with blows of grenades and blows of bayonets drove them out of the city'. Some of the bayonets may have been of the plug-type and others would have been of newer ring-type, which was gaining preference.

When King Charles II died in 1685 his brother succeeded him and was crowned King James II, but he held Roman Catholic religious beliefs which were not popular in a predominantly Protestant England. This led to plots to overthrow him. The first and most serious came in June that year, only four months after he had acceded to the throne in February. His nephew the Duke of Monmouth, although illegitimate, believed he had a claim to the throne and raised a force to invade England. He landed with a small force at Lyme Bay in Dorset and set about raising a force. This was the same Duke of Monmouth who, eight years earlier, had ordered a new type of bayonet to be made. The invasion was abortive and collapsed after Monmouth was defeated at the Battle of Sedgemoor in Somerset on 6 July 1685. One of the leading figures in defending James' monarchy during the emergency was John Churchill, later to become the Duke of Marlborough. This officer had served as an ensign with the guards and in 1667 saw service in Tangiers. During the reign of Charles II he became popular and proved himself a capable commander leading to his being appointed to the rank of colonel in 1678. His popularity continued during the reign of James II, whom he served diligently, and was rewarded with a peerage and made a colonel in the Royal Dragoons.

James II tried to continue his late brother's legacy to create more regiments and in June 1685 he ordered the first official fusilier regiment to be raised directly in England with royal assent. This was the City of London Regiment, which the king referred to as; 'My Royal Regiment of Fusiliers' and issued a royal warrant in which it was stated that: 'Our Royal Regiment of Fusiliers [are] to have snaphance muskets strapt, with bright barrels of three feet eight inches long, with good swords, cartouch-boxes, and bionetts'. The regiment was created by Lord Dartmouth from two companies of Tower Guards in London as the Ordnance Regiment or Royal Regiment of Fusiliers and, as the royal warrant stated, it was armed with flintlock muskets. It was the task of this regiment to march alongside and guard the artillery train, in the same manner as its French counterpart. In the same year grenadier companies in the English army were armed with fuzil flintlock muskets and equipped with bayonets, but they did not carry swords nor any type of shaft weapon. At this time captains had been carrying spontoons measuring eight feet in length as a mark of their rank. Lieutenants carried partisans and sergeants halberds. By 1690 attitudes had changed and grenadier companies were carrying swords and a drill manual entitled *Exercise of Foot* published that year recorded that Grenadiers carried firelock muskets, bayonets, hatchets and grenades. These weapons would be carried as standard by troops serving in these units during the War of the Spanish Succession between France and England (1701–1714). For example, at the Battle of Speyerbach on 15 November 1703 between the

German state of Hesse-Kassel and France, the French commander Marshal Tallard ordered his men to charge at point of bayonet the German troops as they advanced to cross the River Speyer. Tallard's men caught the German troops at a vulnerable moment and after further heavy fighting the French gained a victory, killing and wounding 6,000 men and capturing a further 2,500 for a loss of 3,500 killed and wounded.

In England at this time it was not uncommon for soldiers to be billeted in civilian households, public houses or anywhere else that offered room to sleep. There were no barracks for the accommodation of troops, who could wander into nearby towns wearing their uniforms and carrying side arms, which is to say swords or bayonets. It was not an ideal situation and with little else to occupy them when not on duty, the troops frequented alehouses where they frequently became drunk. This often resulted in a fight with civilians and knives were often drawn and used all too readily. The troops would respond by drawing their bayonets, and serious wounding and even death would result with so many weapons. This matter came to the attention of the military and a royal proclamation on the subject was issued in March 1687, which read:

'For the prevention of mischief that may happen by the carrying of bayonets We hereby strictly forbid all officers and soldiers of quality soever within Our pay or entertainment to carry a dagger or bayonet at any other time than when such officer or soldier shall be upon duty or under their arms upon pain of being punished at Direction of a Court Martial and the officers and commanders in chief of Our several regiments, troops and companies and Governors of Our Garrisons are hereby required to cause these Our commands to be forthwith read and published at the head of each respective regiment, troop and company that all persons may give obedience thereunto.

'Given at Our Court at Whitehall the 4th day of March, 1686/7
  'By His Majesties Command'

The historian Colonel H.C.B. Rogers believed the order was almost certainly drafted personally by King James II, because it carries the same tone in the wording as letters he is known to have written personally on the subject of various military matters. It certainly was a move to try and prevent further enmity between civilian and soldiers when in close contact and, although well intentioned, there would have been those who ignored the order and carried either a dagger or bayonet when going to town.

Three years after the Monmouth Rebellion, King James II was faced by another threat to his monarchy. This time it was his son-in-law the protestant Prince William of Orange and his wife Princess Anne, the daughter of King James II. They landed in Torbay in Devon in November 1688 and their reception was much different to Monmouth's ill-fated invasion. King James believed he could rely on Churchill once again and sent him to confront the landings. Churchill saw which way things were going with a Catholic king and, rather than attack, he took sides with William in the 'Glorious Revolution'. There was nothing left but for James to flee and after a few minor skirmishes, he eventually escaped to France leaving England with a new Protestant monarch. Another new

Wooden-handled plug bayonets were jammed into the barrel of the musket.

Fusilier bodyguard regiment for the artillery was raised, known at the time as Sir Richard Peyton's Regiment of Foot; it would later undergo a series of changes in its title before becoming The Lancashire Fusiliers. Over the years to come, several more fusilier regiments would be raised in England, thereby continuing the trend which had been set in Europe. The title fusiliers is understood to be derived from the type of weapon they carried, the fusil, which had a flintlock action. The fusilier regiments developed a distinct uniform, which would distinguish them as serving in a specialized role. Later, however, military trends would eventually see a decline in the role of such specialized troops on the battlefield as artillery became more autonomous. In recognition of his support, King William showed his gratitude by conferring on Churchill the rank of Lieutenant General and creating him the Earl of Marlborough in 1689. England would now find itself involved in European affairs again, and this put it on a course to go to war with many of them, including the War of the Spanish Succession (1701–1714), throughout which the Duke of Marlborough would campaign.

In February, only three months after their landing, William and Queen Anne found themselves facing the first major trouble confronting their reign. It came in the form of a Jacobite uprising in Scotland where the Highland clans were loyal to King James II, who was a Stuart. After all, his grandfather had been James VI of Scotland and James I of England, and following on from this they regarded him as King James VII of Scotland and gave him their support. Scotland was in turmoil and there was a call for a return to the old ways and the Highlanders loyal to their monarch were prepared to take up arms in the cause. The Lowlanders supported King William, even though he was actually Dutch. It was obvious that things could not continue in such a manner and it was inevitable that it would lead to a clash of arms. Finally, a Scottish army of 2,400 led by John Graham of Claverhouse, 1st Viscount of Dundee, completed marching a route which avoided clashing with the English in order that they should meet on a site of Dundee's choosing. The English army of 3,500 was commanded by General Hugh Mackay who had experience of campaigning from the Dutch Wars. The two sides finally met at Killiecrankie on 27 July 1689. Dundee believed that the site of the battlefield was: 'fair enough to receive the enemy but not to attack them', and despite recognizing that he was outnumbered, he nevertheless ordered his troops to open fire. The English responded and firing continued until finally Dundee decided that enough was enough; shortly before nightfall he ignored his own assessment of the battlefield not being suitable for attack, and ordered his men to advance and charge the English lines. The reputation of the Scottish Highlanders' charge was legendary. It was the suddenness and ferocity combined with the weight and speed with which it was delivered, that made it so effective. The English troops had just fired their muskets and had not yet had time to reload and were in the act of fitting their bayonets ready to receive the charge, when the Scottish troops crashed into them. With unloaded weapons and bayonets not fixed, they were unable to adequately defend themselves. The Highlanders hacked and slashed at close-quarter hand-to-hand combat, and MacKay's remaining men fled the field. Dundee was killed and the Highlanders lost 600 killed and wounded in total. MacKay lost 2,000 killed and wounded, but it was a short-lived victory for the Highlanders. The following month at the Battle of Dunkeld on 21 August the Jacobite Rebellion was effectively broken and the Highlanders scattered and pursued.

In the aftermath of the disaster of Killiecrankie, an assessment was made of what had gone wrong. General MacKay believed that it was partly due to the design of bayonet, which could not be fitted. Some sources claim the English troops had not fitted their bayonets and others claim the bayonets had been fitted but then removed to allow the muskets to be reloaded. In view of what had happened, it is possible that some troops became confused in the heat of battle and panicked as the Highlanders closed in with their swords, and they may have fumbled to either fix or unfix bayonets and try to reload their muskets. It was possible to charge the enemy with plug bayonets fitted, but only when ring bayonets and later socket bayonets were developed, could such a tactic be seen through to a successful conclusion. When that happened troops would be able to advance, only pausing long enough to fire their muskets before resuming their advance and follow up the shock of the volley by pushing home with the point of bayonet.

MacKay later went on to personally claim to have invented a new ring-type bayonet because of his troops fumbling during the battle, but this is questionable because the ring bayonet is known to have been developed in France eleven years before the Battle of Killicrankie. MacKay wrote of his alleged invention in his memoirs, *War Carried on in Scotland and Ireland 1689–1691*: 'All our officers and souldiers were strangers to the Highlanders way of fighting and embattailling, which mainly occasioned the consternations many of them were in; which, to remedy for the ensuing year, having taken notice on this occasion that the Highlanders are of such quick motion, that if a battalion keep up his fire till they be near to make sure of them, they are upon it before our men can come to their second defence, which is the bayonet in the musle [muzzle] of the musket. I say, the General having observed the method of the enemy, he invented the way to fasten the bayonet so to the musle without, by two rings, that the souldiers may safely keep their fire till they pour it into their breasts, and then have no other motion to make but to push as with a pick'. This puts MacKay in the list of those others also laying claim to have invented the design. Among those who dispute his claim to inventing the ring bayonet is the historian Charles Ffoulkes, who also reminds us that the design first appeared in 1678, a fact which is chronicled from the time. What Mackay may have developed was an improved fastener to better secure the bayonet to the musket, which, even then, is giving him the benefit of the doubt, but with little other evidence in the form of something tangible to support his claim, then one is left with no option (despite his writings) but to dismiss his alleged contribution to developing the ring bayonet.

Chapter 4

Time for Change

Like all transitions from old to new, there is a period when both are in use together and the change in design for bayonets from plug-type to ring-type was to prove no different. Both the old and new types continued to be used together and the same thing would happen again when the socket bayonet design was introduced. English bayonets of the plug-type were still being produced at the end of the seventeenth century and some had 'wavy-type' blades or flamboyant blades. The style had been used on swords in the fourteenth century and sometimes referred to as 'flamberge-style' or 'flame-style'. The undulating curves actually produced a longer cutting edge than the traditional straight blade without any increase in the overall length. Some European bladesmiths experimented in producing bayonets that could be transformed from a standard single-blade design into a multi-blade weapon by means of a release catch usually located on the ricasso part of the blade near the handle. A surviving example held in the Wallace Collection in London is of European origin, dated circa 1685, and measures 19 inches in length and weighs less than 1lb. Depressing the catch with the thumb unlocked a pair of spring-loaded side blades, which flicked out to produce a triple-bladed weapon; this turned the musket and bayonet combination into a trident.

Such designs to produce multi-bladed weapons were nothing new and daggers with this kind of action had been produced during the sixteenth century and possibly even earlier. The purpose of having such a weapon was so that the side blades could trap or deflect an opponent's blade. The theory was sound and the design was good; but on the actual battlefield such a contrivance would have had only novelty value, or at the most very limited usefulness. To begin with, the musketeer would have to remember to press the catch to release the side blades. In the heat of battle it would not always be possible when fighting at close quarters and a man could forget. Furthermore, producing such items would have taken longer and the cost due to the additional work involved would have rendered them far more expensive than the ordinary single-blade bayonet; so armies preferred to stay with more practical designs. Some of these triple-bladed bayonets may have been used by select units but there appears to be no record of them having been used in battle. Those surviving examples are today museum curiosities along with hand-axes and swords which incorporate pistols in the handles.

Basic firearms had been in use since the fourteenth century and over the years had evolved into matchlock muskets, which could be fired from the shoulder with the aid of a forked rest to support the weight and length of the fore end of the weapon. Bladed weapons were in abundance along with polearms such as pikes, halberds and Swiss 'ahlspiess' and infantry also carried daggers, but it is incredible to think that it did not occur to anyone to attach some type of blade to the end of the handguns at an earlier time. The weight and cumbersome size of these early muskets may have prevented the combination, but when the first simple plug-type bayonet appeared it was a turning point for the infantry in battlefield tactics.

The use of muskets spread but even then only the best trained and most disciplined musketeers were capable of firing no more than two rounds per minute. Firing produced clouds of thick smoke, which could remain suspended about the positions of the musketeers if there was no wind to disperse it. Further firing produced more dense clouds of smoke which choked the men and stung their eyes and, being unable to see their targets or reload, they were forced by conditions to stop firing, at which point they would find themselves vulnerable. The crucial moment came when they stopped firing and were left exposed to attack by cavalry when they began the process of reloading. That was where the benefit of having a bayonet became useful. At this point an officer commanding a unit would order the fitting of bayonets in a defensive move. The line of musketeers was then transformed into a line of sharp spikes which horses were naturally reluctant to move against. The average length of a musket in the seventeenth century was around five feet and when a bayonet with a blade length of 12 inches was inserted into the barrel, it meant the infantry could lunge forward around six feet to stab either horses or their riders. Even so, it took considerable skill to hold the heavy weapon and wield it with good effect. Indeed, it may have been this waving effect which deterred horses from trying to penetrate.

In effect, the infantry was now capable of being turned into a pike block for their own defence against cavalry. The weapons were a great deal shorter than the traditional pike, which had lengths between 16 and 18 feet, but being shorter meant the infantry were more mobile with this new weapon which could be wielded better. Pike blocks were traditionally the main method of dealing with cavalry, but towards the end of the seventeenth century the pike blocks were becoming fewer and their role diminishing as tactics on the battlefield changed. Even before the outbreak of the English Civil War in 1642 some military leaders were already considering the possibility of abandoning the pike altogether. The pikemen travelled slower because of the weight of their equipment, and there were occasions when they were left behind on the march to enable the rest of the army to move more quickly. They were becoming seen more as a hindrance than an asset as battlefield tactics changed and the style of warfare changed with ever-more emphasis on manoeuvre and speed.

Bladed weapons such as daggers, knives and swords along with pole-arms including spears and halberds had been used by armies for centuries before the advent of the bayonet. Through the use of such weapons at close quarters it had come to be realized that a stabbing action often resulted in killing one's opponent if thrust with sufficient force into a vulnerable part of the body, which was often the throat or abdomen. Despite the portrayal in Hollywood films, such wounds did not always result in an instantaneous death. A man stabbed with a dagger, sword of bayonet could linger for several minutes writhing while he bled to death, but he would invariably be disabled when wounded in such a manner. Slashing actions, on the other hand, may incapacitate an enemy by cutting tendons or muscle, but unless a major artery was severed such a wound rarely produced a fatality. That is not to say such a wound may not become infected, leading to death at a later point. But what is needed on the battlefield is something to produce casualties on the enemy and weaken their resolve to stand and fight – and that something is the bayonet.

The threat from cavalry was universally recognized and the main reason why armies maintained strong blocks of pikemen was to break up mounted charges and protect the musketeers against attack whilst they were reloading their weapons. Whilst pikemen were effective against cavalry, the role absorbed large numbers of troops who could have been put to better use as musketeers. That way, at least, they would have been contributing directly in the battle instead of waiting for the opportune moment when they could be ordered into the fight. The infantry in all armies at the time were armed with the matchlock musket, which was slow and cumbersome to load. In some armies there were thirty movements or more connected with the handling, loading and firing of muskets, and fitting bayonets to these would have only made things more complicated. For example, an official drill book of the period, *An abridgement of the English Military Discipline*, published in 1685, lists thirty movements for loading the flintlock musket and thirty-two movements for loading the matchlock musket. During the time they were reloading, the musketeers were vulnerable to attack by cavalry. This weakness had long been recognized by many military commanders, and a series of tactics were developed to protect them as best as possible on the battlefield. One experimental trial designed to help the musketeers in their own defence was observed by William Barriffe, who recorded in work of 1639, *Militarie Discipline: or the Young Artillery-Man*: 'Having considered the danger of the musketeer and how he is unable to resist the horse after he hath poured forth his piece without hee bee sheltered either by some natural or artificial defence... divers captains and souldiers have often been trying conclusions to make the muskettier as well defensive as offensive. Some by unscrewing the heads of their rests and then screwing their rests into the muzzle of the musket.' This was one method of providing the musketeer with a means of self-defence, but it was far from being an altogether ideal solution as Barriffe later points out in the same writing by stating it: 'proved to be tedious and troublesome.' Still, something was better than nothing at all; and rather an extemporized attempt at producing a pike than risk being exposed to the fury of an attack by cavalry. Without realizing it, the troops engaging in such practices were copying the early plug bayonets by converting a firearm into a thrusting weapon.

The method of using musket rests to provide the musketeers with their own defence to keep cavalry at bay had been mentioned by Gervase Markham in his work *Souldier's Accidence* in 1625 where he comments how the musketeers: 'shall have rests of ash with iron pikes at either end and half hoops of iron about to rest the musket on.' These were stout enough to be used as defence against cavalry and some rests could be connected together by lengths of chain, but this tactic did not prove altogether popular and was rarely used. During his campaign in Egypt in 1796 Napoleon Bonaparte ordered that pikes of five feet in length be issued, and these should be connected together by lengths of chain to provide defence against cavalry attacks. Like the earlier usage of this equipment, it was considered bulky and never popular with troops, who deliberately mislaid the items and the idea was abandoned. Some designs of musket rests were actually developed so that one branch of the 'U-shaped' forked rest was elongated into a spike, which the musketeer could use as protection against cavalry The weapon historian R.J. Wilkinson-Latham claims this device was popular with the French, who he believes developed the item. This form of design was still being made as late as 1700, but as muskets were becoming shorter and lighter this device would have no further use and was eventually phased out of service. Musketeers had long used the tactic of reversing their weapons to use them in a clubbing effect after they had been fired if there was no time to reload before receiving an attack by cavalry. The tactic was crude but it worked, and troops would continue to use it as a last resort in future wars even when the spread of the bayonet in Europe was beginning to make commanders change their tactics. In England such changes would not come about until much later, when Sir James Turner observed in 1670 that: 'Our present militia acknowledging no other weapon for the light armed infantry, but the sword and musket, and this last I have seen sometimes laid aside for a time, that it might not impede the manageing the musket by its embaraas. And indeed when musketeers have spent their powder, and come to blows, the butt-end of their muskets do an enemy more hurt than despicable swords, which most musketeers wear at their sides. In such medleys knives one foot long (the haft made to fit the bore of the musket) will do more execution than either sword or butt of musket'. From the tone of Turner's comments on the use of the bayonet, he sounds as though he was familiar with the bayonet and its use and saw little use for the sword, which he believed was an encumbrance when an infantryman carried a bayonet.

The introduction of the bayonet certainly changed the way the infantry fought; and although it replaced the need to reverse the musket and use it as a club, musketeers would continue this practice as a natural, almost primeval, reaction to danger. The infantry could now advance as a single unit with bayonets fixed in order to stab and slash at the enemy. The pike had now been eclipsed on the battlefield by this new, very simple weapon. But even so, some old practices remained entrenched as Louis de Gaya observed when he noted the methods of the English musketeer and the equipment he carried, but does not mention the bayonet, which had presumably still not entered wide service. He did, however, remark on the use of the butt like a club and wrote: 'quand ils ont fait la decharge du Mousquet ils se batten à Coups de Crosse,' which tells how after firing the musket the butt is used to deliver a 'stroke'. Indeed, in 1678 he went on record by noting that English musketeers rarely drew their swords but still preferred to use the butts of their muskets as clubs. The force used when plying the muskets as clubs was considerable, and the impact could damage the weapon or even break it. In 1670 a unit of Coldstream Guards was engaged in an anti-Puritan operation in Southwark during the course of which they broke 67 muskets when used as clubs, and they also broke 27 pikes in the same operation. Musketeers at this time were also issued with swords but these were never popular with the troops, who believed the long blades suspended from their belts and dangled by their sides impeded their free movement. The sword was obviously useful but with the introduction of the bayonet with its much shorter blade, it came to be used more widely and the sword was gradually discarded by foot troops.

During the early part of the sixteenth century pike blocks had formed as much as 80 per cent of infantry units on a battlefield, but gradually as muskets became more widely used the men to fire them were taken from the pike block formations. In Europe by the mid-seventeenth century the pike was falling out of use in many armies, but in England where the Civil War raged it was still important in the armies of both sides. After 1660 the English army underwent changes and as a result the pike gradually disappeared, but some last vestiges of it would remain as the NCO's short staff known as a spontoon. The length of the pike varied also during this period. In 1645 the average length in some armies was around 15 feet, but could be as much as 18 feet. By the early 1700s they were reduced in length to around 14 feet. The numbers of pikes being deployed on the battlefield by around 1697 after the Peace of Ryswick had fallen from a ratio of 1:3 in some armies to 1:4, and the decline continued as the pike was substituted for the musket and bayonet. It was the subject of a debate which would last for around twenty-five years, with many arguing that with the continuing widespread use of the musket, there was hardly a reason for the pike on the battlefield. Indeed, even Louvois commented on the argument surrounding the replacement of the pike by the bayonet and wrote how: 'The substitution of the pike with the bayonet was the subject of a fight which lasted twenty-five years. Since the use of the gunpowder, the pike hardly had a reason to be. Fourteen feet long, difficult to handle, in battle this heavy weapon immobilized two thirds of the most vigorous men of the company'. (The Marquis de Louvois served as the French Secretary of State for War during the reign of King Louis XIV.) He also wrote to his contemporary, Vauban, how some German states had given up on use of the pike and even the Hungarian army had also discarded the pike.

Eventually by the time the War of the Spanish Succession, (1701–1714) had been concluded, the pike had all but been replaced in most European armies and the bayonet had been in service for more than 50 years. Jacques de Chastenet noted this transition in his work *Art de la Guerre*, remarking how during the War of Spanish Succession: 'When the war started, there was already some regiments which had quit using pikes. The remainder always had a fifth of the soldiers armed with pikes; but by the winter of 1703–1704 they were entirely given up for muskets shortly afterwards. During this war the officers were armed with spontoons of eight feet length; sergants of halberds of six and a half, and all the soldiers with muskets, with bayonets with sockets, to be able to fire with the bayonet at the end of muskets'. At this time the English army is known to have been using both plug and socket bayonet designs together, because accounts show that 3,000 socket bayonets were taken from storage in the Tower of London and sent to troops on campaign in Portugal, along with a number of socket bayonets. Documents from this period also give us the date when pikes were abandoned in the English army from a Board of Ordnance record dated 1706, which states how: 'All the regiments raised since the disuse of pikes (1702) have provided bayonets… at their own charge. Few of the officers agree in the sort of bayonets fit to be used or in the manner of fixing them as may appear by the various sorts there are of them in the Army.' This tells us there were a number of different types of bayonets and fixings in service which must have caused a number of problems. The year after this document appeared the English army became known as the British army following the Act of Union in 1707 between England and Scotland. English troops had already been using bayonets since around 1663 when they were first issued to the 2nd Regiment of Foot (later to become The Queens (Royal West Surrey Regiment) raised in 1661) when the regiment was posted in Tangier.

An observation made on the transition to bayonets and the discarding of pikes was made by General Fuller, who wrote how the change: 'revolutionised infantry tactics. First, it reduced infantry from two types to one and simplified fighting; secondly, it enabled infantry to reload under cover of their bayonets; thirdly, to face cavalry; and fourthly, to protect themselves in wet and windy weather when firing was restricted.' He continued by stating that: 'this radical change in armament coincided with a radical change in the outlook on war itself.' Indeed, from this point onwards there would be further developments in tactics and strategies on the battlefield.

By 1671 the plug bayonet had been in service for more than twenty years and although undoubtedly useful its design was not ideal. The long, sharp blade was not the problem, but rather it was the method of plugging it into the barrel. A new design was what was needed and during the reign of King Louis XIV of France an improved type was developed. The bayonet was, by now, known about across much of Europe and although used by some troops, in some quarters it was still a relatively unknown and new weapon. When the ring bayonet design came to the attention of the French king he ordered that it be introduced throughout his army as a standard issue item. The troops carried the bayonet in a scabbard sometimes called a 'sleeve', which indicates that it could be carried suspended from a belt and ready to hand for use. It would appear from various descriptions, such as that written by Sir James Turner in the same year, that the bayonet was becoming more uniform in style and tells how the blades are 12 inches in length. The bayonet would traditionally be worn on the waist belt where it was ready to hand for soldiers to reach and fit quickly and easily. The American Army broke with tradition and in the twentieth century during both world wars for some reason decided that in combat the bayonet would be carried attached to the troops' rucksacks and this positioning would cause problems in reaching it.

The Swedish army was using plug bayonets before the end of the seventeenth century and went on to produce some types with blades of exceptional length, which could be used as swords in an emergency. The country also adopted the ring bayonet design, but around 1692 a small but significant change was made in the basic design by incorporating a small spring clip to the handle, which gave it a more secure fix when inserted into a pair of rings mounted alongside the barrel. The problem of developing a secure fixing for the ring type of bayonet appears to have been a matter of some concern in a number of armies. For example, on 21 August 1721 Isaac de Chaumette took out patent number 434 for 'Swords to serve as bayonets by means of a ring at the pommel and a screw at the hilt', with the British Patent Office. Unfortunately, no further details are given or they have been lost over time. These were but two variations on the ring bayonet design, which had been developed in France around 1678 as noted by Puységur. The ring bayonet design allowed troops to fire their muskets with bayonets fitted but, with the blade lying straight along the barrel, it still proved an obstruction during reloading. On firing the musket ball could strike the blade and if this happened, the bayonet could become damaged or broken. Muskets were not inherently accurate to begin with and, had a musket ball struck the blade of the bayonet, it might be deflected, thereby degrading the accuracy of the weapon even further.

Nevertheless, it was another step forward in perfecting the optimum design of the bayonet, which was evolving at different rates from one country to another. In England, for example, bladesmiths were still producing plug-type bayonets as late as 1695 and some designs had blades measuring a formidable 24 inches and weighing over 1.5lbs to present an extremely intimidating weapon at the end of a musket. Unfortunately, the longer the blade the greater the chances were of it twisting when being thrust into an enemy. To reduce this effect, a special rib or raised edge running along the length of the blade was included in the design to give rigidity. Such extended lengths would have made controlling the musket and blade quite difficult, and these types of bayonets would have been more useful being held in the hand as a short sword is in close-quarter fighting. Other countries, meanwhile, in particular France, were beginning to develop the first early designs which would evolve into the socket bayonet.

The length of bayonet blades always exceeded the minimum that was known to be required to kill or incapacitate. Blade lengths of 15 to 18 inches, or even longer, were more than sufficient to pierce an average-sized man's torso from front to back and even protrude. Such blade lengths could penetrate deep into a horse's abdomen to bring it down, which is the primary purpose of the bayonet in providing a defensive measure against cavalry attack. It was a well-known and established fact that a puncture wound of only a few inches depth was all that was required to kill with a stabbing thrust, but bayonets were always produced with very long blades. The blade being longer than necessary to complete the task is what is termed as 'overkill' in modern usage. It was the great length which made the bayonet so intimidating, and armies capitalized on this fact before realizing that there was an optimum length which a design could reach before becoming an encumbrance. A shorter blade could achieve the same result and modern designs are therefore much shorter while still retaining the same psychological effect, although in the nineteenth and early part of the twentieth centuries this was not considered to be the case. An adversary may not know if the musket of his opponent was loaded and ready to fire, but the long-bladed bayonet fixed to the end of barrel would have left him in no doubt as to what it was intended for, and that was to kill. No man on the battlefield ever wants to be killed and never thinks it will happen to him in this most painful manner, by being stabbed and bleeding to death.

By about 1700 the ring bayonet was still being used by the English army, as noted by the historian F. Grose, who mentions it in his work *Military*

Antiquities of 1801, where he recounts the observations made by a certain Reverend Gostling from Canterbury in Kent. Apparently, the good reverend saw two horse grenadiers riding as escort to the coach of Queen Anne with muskets to which were attached 'bayonets fixed by means of rings.' An illustration was provided by Reverend Gostling which appeared in Grose's work, but supporting details to confirm the accuracy of the image were not supplied. Indeed, from the sketch the bayonet resembles a standard plug bayonet with rings attached, which could slide the length of the handle. In his *Memoirs* published in 1737, the Marquis de Feuquières mentions: 'Bayonet: a short, broad dagger made with iron handles and rings that go over the muzzle of the firelock.' The problem with this arrangement was that not all barrels of muskets were made to a standard size and there were variations in the diameter of the barrels. This meant that sometimes the rings were too small to slide over the barrel and some were too large to provide a firm and secure fitting. An alternative to fitting the rings to the handles of the bayonets was to fit the rings to the barrels of the muskets, which allowed the continued use of plug bayonets. In the space of only thirty years the design of the bayonet had undergone a change in how it was fitted to the musket and another change would come before the end of the century; and this development, with some modifications, would remain in universal service until the mid-nineteenth century. Despite all these changes, the plug-type style of bayonet remained in use with hunters in Spain until the nineteenth century, by which time only civilians found they still required the simple method of pushing it into the muzzle of the barrel to produce a spear to hunt their quarry which included wild boar.

So the question is: What, after all this time, is a bayonet, how is it defined and what places it apart from knives? The answer to this is perhaps best given by the following description: 'A bayonet is a blade that can be attached to the muzzle of a gun. In its original and most important role, it converts an infantryman's musket or rifle into a pike'. The operative word in this description is 'blade' because over the centuries the bayonet has appeared in a wide range of shapes and sizes, but always with a fixed blade. The bayonet certainly began as a hunting knife and in 1660 the writer Favè records in his work, *Études sur le passé et l'avenir de l'Artillerie*, following a proclamation by King Louis XIV, commenting: 'La frequence des accidents qui arrivent journellement par usage des baionnettes et couteaux… qui se mettent au bout des fusils de chasse ou se portent dans la poche.' This tells us how the discovery of bayonets came about by accident or chance when it was found that a hunting knife or some type of dagger, which is usually carried in the pocket, could be fitted to the end of a gun and used during the hunt.

This discovery was commented on by Chevalier Jacques de Chastenet, Seigneur de Puységur, who wrote how in his opinion that a 'couteau de chasse' (hunting knife) was better than a sword. For close-quarter fighting in a melee that was certainly the case, because it could be used in a short stabbing action. Puységur's main work *Art de la Guerre* was published posthumously by his son in 1748, and contains at one point a criticism concerning the 'baionnette à douille' an early form of socket fixing for the bayonet as devised by Sébastien Le Prestre, Seigneur de Vauban (later being made Marquis de Vauban and created a marshal). A version of the term 'douille', which translates as 'socket', also appeared in the work entitled *Histoire de la Milice Francaise*, which was written by the French historian Gabriel Daniel (1649–1728), and appeared in 1724. In that same year Humphrey Bland published his *Treatise of Military Discipline*, which was an instruction manual and one of the directives it contained was that troops should: 'Charge your bayonets chest high.' This was a stance which would be used by the Duke of Cumberland's troops at the Battle of Culloden in 1745. Vauban was a French military engineer, born in 1633, and devised new methods of constructing fortifications, developed improved methods of conducting siege warfare, and was interested in many other things. This association with all things connected with the military led to him devising an improved socket bayonet to replace the earlier plug- and ring-type bayonets, and he is understood to have been influential in recommending that the French army adopt the flintlock musket. He served on campaign and was even wounded and his ideas influenced others, such as the Dutch military engineer Menno Van Coehoorn (1641–1704). French troops serving in Canada were being issued with socket bayonets, which they fitted to the Model 1717 musket, and the use continued with the Model 1728 musket which replaced the earlier plug-type.

The idea for a new type of bayonet as devised by Vauban was to produce a design which had a socket-like sleeve to which was attached the blade and which would allow it to: 'encircle the muzzle [of the barrel]… and leave the musket perfectly free for firing.' In his work Puységur wrote how: 'During the war of 1688, one proposal made to the king [Louis XIV of France] to discard pikes and give pikemen muskets. He [Vauban] made a demonstration of bayonets with sockets on the muskets of the regiment. The bayonets had not been made with the muskets, which were of different sizes, and did not fit firmly. In the presence of his majesty, several bayonets were made to fall off while withdrawing and others when the bullet was leaving the muzzle, so they were rejected. Shortly afterwards, nations with which we were at war discarded pikes for muskets with bayonets, while we were obliged to use pikes.' The war to which Puységur was referring was the Nine Years' War of 1688 to 1697 and his observation that bayonets fell off 'while withdrawing' is obviously a reference to withdrawing the ramrod during the loading process of the musket.

Undeterred by the poor showing of his new design for fixing the bayonet to the musket, Vauban went back to the drawing board and revised his design and made changes to correct the fault. One of the changes he made to the design was to fit the bayonet with a small clip to produce a firm fitting. Vauban wrote a letter to the Marquis de Louvois, François Michel le Tellier, the Secretary of State for War, in which he stated that he believed his design could and would work if given the chance to prove it. He went on to assert that he believed troops armed with the new bayonet fitted to their muskets and placed in front of formations could defend the unit against attack by cavalry. Louvois was one of King Louis XIV's leading ministers and had been responsible for centralizing the army's administration and he would improve equipment, which also included replacing the matchlock with the better flintlock musket and the introduction of the socket bayonet, which had been brought to his attention by Vauban.

The most commonly used musket by all armies at the time was a weapon which was fired by a method of operation known as 'matchlock'. These were heavy, bulky weapons often measuring five feet in length, and some were so large and heavy they had to be fired using a forked rest to balance the fore-stock. Longer muskets known as 'fowling' pieces were civilian weapons which were originally used for hunting birds, but because of their accuracy they were used for long-range shooting or sniping against officers and drummers who were responsible for relaying signals. These weapons could be six feet in length which meant that they too, like the standard military musket, were not ideally suited to bayonets being fitted. The length would have given the musketeers greater reach when lunging at cavalry but the bulk and weight precluded them from being used in such a way. Another type of musket in limited use at the time was the 'flintlock' but these were more expensive and reserved for use by elite units such as Guards.

The flintlock was loaded in the same way as the matchlock, which is to say a measured amount of powder was tipped into the barrel and rammed down using a wooden ramrod. A lead ball was then rammed down on top of the powder. The frizzen pan was primed with powder and the weapon was ready to be fired. In the case of the matchlock this was done by pulling the trigger which operated a lever to lower a length of burning cord known as a slow match onto the powder in the frizzen pan and the weapon was fired. The flintlock used a piece of specially-shaped flint held in a small vice-like device known as the 'dog', and when the firer pulled the trigger, the spring-loaded mechanism allowed the device holding the flint to snap forward to strike a steel plate creating sparks; this ignited the powder in the frizzen pan, and the weapon fired. The propellant powder often came as a prepared charge wrapped in a paper tube, known as the cartridge, which made it easier to load into the barrel. The flintlock operation had been in use for some time but it was an expensive weapon to produce and this had prevented it from being widely used by armies. At first the flintlock was used by specialist troops such as fusiliers but eventually the weapon's more reliable method of operation was recognized as being an improvement on the matchlock and from around 1620 the flintlock musket spread further. The flintlock musket, being smaller than its predecessor the matchlock, was better suited to being fitted with the bayonet. These weapons could accept either plug or ring bayonets and later would also allow the early style of socket bayonets to be fitted with some modifications.

The main advantage with the ring bayonet when it was attached to the musket was that it allowed the infantry to fire their weapons with the bayonet fixed whilst advancing. To troops facing their adversaries with bayonets fitted, they would see that the bayonets were fitted and would naturally assume the muskets were not loaded, especially if they were unaware of the ring-type fitting. It would have come as a terrible shock to them to receive a volley of musket fire, which was then followed up by the 'point of the bayonet'. The effect would have been quite staggering to be fired at and, before they could compose themselves or recover their position, the approaching enemy would have moved closer to press home the attack with the point of bayonet. It was one of the first tactics to be specially developed for the use of the bayonet, and now meant the bayonet could be used offensively as well as defensively in conjunction with a musket which could always be fired.

The ring bayonet design was a natural progression in moving the development of the bayonet forward and marked the point which turned it into an indispensable infantry weapon. Whilst the ring method of attaching the bayonet to the barrel of a musket was not perfect, the method did work. The main problem lay in the fact that the diameters of the barrels of the muskets varied in size and the rings were either too large or too small to fit the barrel. Some armies used three rings to attach the bayonet, but the most widely used form was the double ring method, and some countries, such as Sweden, fitted spring-loaded clips to the handles of the bayonets to take up the slack if the ring attachments were too large and provide a firmer method of fixing. The next obvious step was to develop a better method of fixing the bayonet to the musket and this appeared in the shape of Vauban's socket design. However, as observers such as Puységur noted, the style was not without its problems due to the differences in the diameter size of the musket barrels. This led to the split socket design being devised. No-one knows who actually came up with the idea originally, but some weapon historians place its date of development to around 1671, which is some several years before the introduction of the ring fitting.

As with other interim stages in changes in design, the split socket arrangement was not perfect but it worked and for that reason can be considered a stop-gap expedient until the true socket bayonet was perfected. The split socket design comprised of a blade fixed to a sleeve or tube, which fitted over the muzzle of a barrel, but to make allowances for differences in sizes the sleeve was split along its length. This meant that the sleeve could be prised open or pinched closed to fit the diameter of the barrel, whatever its size. It was the only way in which the problem of fixing could be overcome and although it was probably meant as a temporary fix, it worked well and surviving examples of the design show that it was still in use well into the eighteenth century with some dated to 1700 and even later. Records from 1706 state that the regiment, which would later become known as the Killigrew Dragoons, were issued with this type of bayonet: 'to serve over a full-bored musket.' The regiment could trace its origins back to 1693 when it was raised as Henry Conyngham's Regiment of Dragoons in Derry, Ireland. In 1704 the regiment fought in the War of the Spanish Succession and in 1707 were renamed Killigrew's Dragoons after their commanding officer Colonel Robert Killigrew. It is possible the discrepancy in the dates could be due to the fact the report was published at the time when the regiment's name was being changed, thereby causing an overlap of the dates. By 1775 the regiment would be renamed yet again and called the 8th Dragoons.

The ring bayonet and some of the early split socket types were designed in such a way that the blade lay in line with the barrel and did not allow for space for a man to complete the movements to reload his weapon freely. To overcome this problem the simplest solution was to 'crank' the blade to create an angle in order to produce a gap between the blade and the barrel of the musket, so that the infantryman could now use his ramrod to load his musket without any impediment. Even with the benefits which the new designs brought, there was still much work to be done and it was far from being entirely successful. Few new ideas work the first time they are tried and, although the socket bayonet was a good idea, the early versions were still prone to working loose during firing. Even the split socket types suffered from this defect because of the variation in the diameter of the barrels. In order to rectify this problem various methods were worked on from the original idea. Among those methods tried in an effort to prevent the bayonet from working loose included fitting small screws and 'locking rings'. It was in Sweden again, around 1696, that one solution was tried, but not, it would appear, copied elsewhere. The solution the Swedish military tried was to fit a small wing screw to the socket so that when screwed in, it tightened against the side of the barrel. It was effective but rather fiddly to operate, especially in cold or wet weather when the soldiers' hands would not be able to manipulate such an item properly and the idea was eventually dropped. The blade was still the flat knife-type design, but some armies such as the French were beginning to experiment with blades which were triangular in section. The French are also credited with developing a new method of style for attaching the blade of the socket bayonet, which now comprised the sleeve into which was cut a 'zig-zag' opening to accommodate a stud on the barrel, and this was called either the 'mortise' or the 'locking slot'. The new layout for the blade was referred to as being 'cranked' to give a bend known as the 'elbow' and then bent forward through 90 degrees. This new style is believed to have originated around 1697 but some sources date it earlier. At the end of the blade closest to the socket sleeve, which had once been the quillons, some bladesmiths were incorporating a small integral ring called the 'shoulder'. The blade would continue to vary in shape and size between armies, and even within the same army, as various bladesmiths experimented with designs. Despite the initial drawbacks, the socket bayonet would prove to be the much better design and with modifications would remain in service until the second half of the nineteenth century.

The bayonet was, and still remains, a simple bladed weapon like any knife and it does nothing until it is wielded in order for it to be useful as a stabbing weapon. Soldiers in armies across Europe were now being issued with bayonets and in doing so, all infantry now had the capability of being offensive and defensive. It was something that appealed to civilians and soon some gunsmiths were responding to the call and developed short spring-loaded blades that were fitted under the barrels of pistols, and which folded back along the length of the weapon. A release mechanism, sometimes operated by pulling the trigger guard to the rear, unlocked the blade and allowed it to snap forward. Such devices whilst sometimes referred to as bayonets, were intended for personal protection and therefore not technically bayonets in the true sense. A range of these types of pistols were made by gunsmiths such as Henry Twigg of London and they became popular in the mid-eighteenth century. The design was intended to allow the pistol to be used as a knife in self-defence after the pistol had been fired. Guards on mail coaches were armed with types of weapons called a blunderbuss to prevent robbery by highwaymen. Some of these designs also had spring-loaded bayonets fitted like the pistols. Gunsmiths such as Manton, Grice, Smart and Waters made weapons of this type from around the eighteenth century. They continued to be made into the nineteenth century by gunsmiths such as Mortimer, but these weapons were for civilian use as protection against robbers and cannot be considered as military weapons. Indeed, in the military such weapons would have been useless, due to their relatively small size, and the bayonets the soldiers used in battle were a far cry from these deadly novelty weapons. Over time, though, designs would improve and become strengthened and the spring-loaded bayonet design would return to be used on military firearms such as the Russian Mosin-Nagant rifle. The Italians expressed a liking for folding bayonets on their weapons and during the Second World War used three main types on the Mannlicher-Carcano rifle along with the Beretta Modello 18 and the Beretta Modello 1918/30, which were both carbines. In post-war years the Chinese-built version of the successful Russian-designed AK-47 assault rifle known as the Type 56, incorporated a folding bayonet and this is still in use with armies around the world today.

It was only right that gunmakers should produce bayonets, especially when they had been developed specifically for use with rifles they had designed. For example, London-based gunmaker Durs Egg developed a bayonet for his carbine which was in use with some mounted troops. What made his bayonet

The so-called 'spear bayonet' designed by the gun-maker Durs Egg, to fit carbines carried by cavalry. It was not successful.

so different was the fact that it was designed to turn the carbine into a spear or short lance for the cavalry. In 1784 Egg was commissioned by the Duke of Richmond as the Master-general of the Ordnance to design a bayonet for cavalry carbines. The result was a most unusual design and certainly not a form one would have expected a bayonet to take. Durs Egg produced a long rod tipped with a leaf-shaped blade, with an example weighing around 1lb and measuring 25.5 inches in length. The spear-bayonet was fitted with a socket attachment for attachment to the carbine. Unfortunately, this movement would not have been easy for a mounted man to complete and there would have been the risk of dropping the bayonet as he tried to control his horse at the same time. The shape and length of the bayonet also meant that it could not be worn on the rider's equipment like his sword. The Durs Egg bayonet, when not fitted for use, was carried reversed under the barrel of the carbine with the tip clipped into a recess by the trigger guard.

Despite it being a clumsy weapon for a cavalryman to handle whilst mounted in a saddle, the 'spear' bayonet was recommended for service and in 1785 it was suggested that the 7th, 10th, 11th, 15th and 16th Light Dragoons be issued with it for use with their carbines. These five regiments all received spear bayonets in 1786. Two years later in May 1788 a report on the serviceability of the bayonet was prepared and concluded by stating that improvements be made to the socket fitting, and it went further by saying that perhaps a 'folding' bayonet would be better suited to mounted troops. The fact that it also appeared in three different lengths did not make it more acceptable to troops. The spear bayonet was removed from service and joined the list of failed designs. It was a most unusual design but others would appear in the future, some of which would be equally useless and others would be practical. Today, examples of these are held in private and museum collections where they still arouse curiosity as oddities.

The introduction of the bayonet may have been a gradual process but its evolution from the plug type of the first half of the seventeenth century to the socket type happened relatively quickly, and led to it becoming part of the ancient steel 'arme blanche' of European armies by the close of the century. Before that time experiments were being made with the shape of the bayonet blade. Increasing the length was an obvious move to increase the reach when lunging at an opponent, and also to produce a deeper penetrating wound. The width of the blade was also increased to produce a wider gaping wound, but the style always returned to the long tapering blade of approximately 12 inches in length as this was considered the optimum to produce a disabling or fatal wound and was manageable for the infantry, who did not need to be burdened with a weapon of unnecessary length. France had led the way from the beginning and maintained this lead by adopting the socket bayonet in 1689. The German State of Brandenburg-Prussia followed this move a year later and then Denmark, which adopted the design in 1690. In that year the French fought an Alliance force of troops from the Dutch Republic, Spanish Empire and the Holy Roman Empire at the Battle of Fleurus, an engagement of the Nine Years' War, where they used socket bayonets with the 'zig-zag' cut into the sleeve to permit the bayonet's attachment to a musket. Prussia would be at the forefront of changes in bayonet design again in 1787 when it adopted the sword-type bayonet, a style which other Europeans countries would copy over the next thirty or forty years. Towards the end of the seventeenth century, weapon manufacturers experimented with the development of a spring-loaded bayonet, which folded back along the underside of the musket, but it was never pursued to full development. Some navies did use the short-barrelled 'blunderbuss-type' weapon fitted with folding bayonet. This was a useful short-range weapon designed to be used in the confined spaces of a ship to repel boarders, and the bayonet was handy after the weapon had been fired. The spring-loaded bayonet was a design which would be resurrected centuries later during the First and Second World Wars and post-war weapons such as the Soviet-designed AK-47 so-called assault rifle.

The first widely accepted bayonet charge in history is credited to French troops who are recorded using bayonets for the specific purpose of engaging the enemy at close quarters at the Battle of Marsaglia, at Orbassano near Turin in Italy on 4 October 1693. Marshal Nicolas Catinat, commanding a French army of 35,000, inflicted a resounding defeat on a Spanish army of 30,000 commanded by Duke Victor Amadeus II of Savoy during the Nine Years' War, also referred to as the War of the Grand Alliance, of 1688 to 1697. The war had a number of causes but it was mainly a territorial dispute between King Louis XIV of France and many other European heads of state. Historians such as Emmanuel Broglie and Rousset record how during the battle the French troops were given the order: 'Les charges, commandèes par les officiers, l'epee à la main, et faietes, d'apres l'ordre exprès du marèchal [Nicolas Catinat] "… au pas de course' la baionette au bout du fusil et sans tirer un coup…" furent remarquables et executes avec ube vigueur qui dècida du sort de la journèe.' This tells us that the commanding officers acting on the express orders of Marshal Catinat ordered the bayonet charge. One account tells how, 'They will order, in their brigades, that the battalions put their bayonets at the end of their musket and do not fire. All the battalion will go in at the same time to enter the line of the enemy.' During the battle the French lost 1,800 killed and wounded, whilst the Italianate Duchy of Savoy and Spanish forces lost 10,000 killed, wounded and captured. The bayonet charge itself was almost certainly not the decisive moment of the battle and was probably conducted in a manner similar to that of infantry armed with spears, which is what the early plug bayonet and musket combination would have resembled. The determined charge would have followed a volley of musket fire from massed ranks, then either plug or ring bayonets would have been fitted to demonstrate a tactic which would be further developed and strengthened over time. It was possible to charge with plug bayonets fitted, but it was only once the socket bayonet had been developed that a combined and bayonet charge tactic could be seen through to a decisive conclusion. Troops equipped with the socket bayonet could now fire their muskets whilst advancing and then follow up the volley with the shock action of the bayonet charge. This was now the beginning of tactics which had been specially developed for the bayonet.

Catinat's victory was gained by superior firepower and the fact that he had deployed his forces skilfully and made use of his greater numbers. The French infantry at Marsaglia are recorded as advancing boldly up to the Italo–Spanish positions without firing a shot. It is claimed the bayonet charge inflicted heavy losses among the enemy and caused a route. After the battle Catinat wrote to King Louis XIV, stating: 'I do not believe, Sire, that there has ever been an action which has shown better what your majesty's infantry is capable of'. The French army at this time still maintained the upper hand when it came to the use of the bayonet, even though other armies had copied the development. They had been the first army and for a time the only force to know about and appreciate its use. The first plug-type bayonets when fitted certainly enabled the troops to turn their muskets into replacement pikes, but there was a time when it looked as though the bayonet when first introduced had a greater role as an accoutrement for parades. Within ten years Catinat's tactics were being copied by commanders such as Tallard, who used them to good effect at the Battle of the Spires in 1703, and the Duc de Vendome at the Battle of Calcinato in Italy on 19 April 1706. Other commanders such as James Wolfe, John Churchill, Duke of Marlborough and the Duke of Cumberland would all come to realize the value of the bayonet as a weapon which could decide a battle at close quarters. According to Colonel Home in some miscellaneous papers, he believed that: 'The introduction of the bayonet marks the end of medieval and the beginning of modern war.' He continued by explaining how: 'Tactics were revolutionised by a dagger 12 inches long' Colonel Home was correct in his assessment of tactics concerning the introduction of the bayonet, because if it were to be optimized, then drills for its use in battle had to be developed and when they were it did change the way battles were fought. With regard to the transition from medieval to modern war, this was going a step too far because warfare had been changing from the moment gunpowder weapons appeared on the battlefield. The bayonet undoubtedly added to this and made its own mark and continued the process of evolution.

There appears to be some discrepancy about Marsaglia being the first battle where a bayonet charge was conducted en masse and some sources, such as Eugene de St Hilaire in his work *Histoire d'Espagne*, give credit for the first bayonet charge to the French Guards three months earlier at the Battle of Neerwinden (also known as either Neerlanden or Landen) on 29 July 1693 where a French army of 80,000, commanded by Marshal Luxembourg, defeated an Allied army of 50,000 English, Scots and Dutch, commanded by King William III of England. By now the bayonet had been in service for around fifty years and would almost have certainly been used in skirmishes or minor engagements. The difference with these battles was the sheer scale and that would have given greater prominence to the role of the bayonet, which would have been more noticeable during the course of the fighting.

The Battle of Neerwinden developed into hand-to-hand fighting and it was almost certain that the bayonet would have been used because conditions were ideal for its use in such circumstances. The French lost 9,000 killed and wounded, but succeeded in driving the Allies from the field and inflicting 19,000 killed, wounded and captured. This phase of the fighting also included the siege of the garrison at the fortified Spanish town of Charleroi by the French. Over the centuries, the bayonet charge would come to prove itself a decisive tactic at the climax of a battle where an army may be beaten but not actually broken. At the sight of the dominant force approaching their positions with muskets tipped with bayonets, the wavering forces would flee rather than face up to the final act of crossing steel. Some officers would espouse the use of the bayonet, such as the Dutch-born General Baron David Hendrick Chassè, whose fondness for the use of the weapon earned him the nickname 'General Bayonet'. He was not the only one and over time there would develop a 'cult of the bayonet', which would continue well into the nineteenth century. By contrast there were some commanders, such as the Swiss-born General, later Baron, Antoine-Henri Jomini who served in the French army of Napoleon Bonaparte, and declared how during his military career: 'I have seen melecs of infantry in defiles and villages, where the heads of columns come in actual collision and thrust at each other with the bayonet; but I never saw such a thing on a regular field of battle'. The historian Jock Haswell believed the bayonet: 'was rather more frightening than lethal because the bayonet charge was usually enough; few troops would stand and close with ranks of men determined to use it.' This is open to conjecture because no one could tell how troops would react in the face of a bayonet charge. In many cases the men would indeed break ranks, but there were also many occasions when they stood fast and fought it out. The historian David French wrote that in his opinion: 'If troops had a sufficiently high morale, infantry attacks could still culminate in a successful bayonet charge.' But as weapons became increasingly more powerful, the price the infantry would have to pay for such tactics would be extremely high, as proven when they faced the firepower of artillery and later machine guns when attacking trenches in the First World War.

Chapter 5

The Eighteenth Century and the American War of Independence

The concept of the socket bayonet may have been an original idea from the genius of Vauban, but it took the combined skills of the gunsmith and bladesmith to make the weapon work. It was an innovative move and advanced the role of the infantryman in battle by returning his versatility once more by looking back to the age of the order when each man could act independently. The bladesmiths in France began to produce the socket design and the gunsmiths had to modify the barrels by standardizing the size of the muskets they were producing and incorporate attachment points. This meant the fitting of a small, fixed 'stud' near the muzzle of the barrel onto which the slot in the socket could be attached.

The ring bayonet design allowed the musket to be fired with it fitted, but the blade lying alongside the barrel caused a problem to the musketeer. When fired the musket ball could strike the blade and damage or break the bayonet. Even if the ball did not damage the blade, the fact of it striking the bayonet at all could affect the accuracy of the musket – not that these early firearms were particularly accurate. This fact can be evidenced by the firing trial conducted by the Prussian army to demonstrate the inaccuracy of the musket in the eighteenth century. There were a number of similar trials conducted, any of which could be used to prove how inaccurate the weapons were. The Prussian army erected a screen of canvas measuring 100 feet wide and six feet in height on a frame, which was meant to represent an advancing unit of enemy infantry. The Prussian infantry, using standard muskets, fired at this target from a range of 225 yards and achieved only 25 per cent hits. The range was closed to 150 yards and 40 per cent of those firing at the target were registered as hitting it.

At the range of only 75 yards the number of those who hit the target increased to 60, thereby confirming that the closer one was to the target, the better chance there was of hitting it. Officers commanding troops in the field noted this too, such as Major George Hangar, who served during the American War of Independence and who in 1805 as a Colonel wrote of his experiences and observations on musketry in the British army at the time: 'A soldier's musket if not exceedingly ill bored (as many of them are) will strike the figure of a man at 80 yards; it may even at 100; but a soldier must be very unfortunate indeed who shall be wounded by a common musket at 150 yards.' At the Battle of Bunker Hill, 17 June 1775, the cry went out for the Colonial troops to 'Hold your fire until you see the whites of their eyes.' This was to maximize firepower and better ensure the chance of hitting a target. It was this inaccuracy which was one of the reasons why the blade was 'cranked' away from the barrel and developed into a style which was copied universally and remained in use until the late nineteenth century. Cranking the blade away from the musket barrel also provided space for a man to insert his hand when reloading the weapon.

During the eighteenth century armies grew in size and the numbers of casualties also increased due to the firepower, range of artillery and improved training in using these weapons. In 1724, during the reign of Tsar Peter the Great, the strength of the Russian Army was maintained at a peacetime level of 170,000 through conscription at a time when the population of the country was around 17 million. At the Battle of Blenheim, 13 August 1704, the victorious Allies lost around 12,500 killed and wounded or around 24 per cent of their forces, which numbered 52,000 troops. The French lost 20,000 killed and wounded or almost 40 per cent of their forces in an engagement which lasted around five hours. Their defeat was made all the more poignant by the capture of a further 14,200 troops and all of these casualties had to be replaced, equipped and trained with everything from trousers, boots and muskets to bayonets. Fifty years later, the Prussian Army would lose 180,000 men killed during the entire Seven Years' War, 1756–1763. This figure was equal to three times the size of the Prussian Army at the beginning of the war. Each man killed had to be replaced, along with those disabled by their wounds, such as the loss of an arm or leg, which prevented them from continuing in any further military service. The Prussian King Frederick the Great was moved to comment on the situation as he saw it by stating that: 'recruits can replace the numbers but not their quality.' He went on to continue his observations by stating that: 'One commands in the end nothing more than a band of badly drilled and badly disciplined rustics'.

Frederick's opinion of the bayonet was unwavering to the point that in some of his early engagements, he ordered the infantry to advance a point of fixed bayonet giving 'fire as little as possible with the infantry in battle; charge with the bayonet.' The high casualty rate due to this suicidal tactic led him to re-examine this approach and develop other methods so that his troops gave the enemy 'a stirring volley in the face. Immediately thereafter they should plunge the bayonet into the enemy's ribs.' This was a tactic used by the British army and would stand in good stead with other armies across Europe. Prussian military discipline was harsh and the bayonet was also used to enforce this. Regulations stated that 'if a soldier during an action looks as if he is about to flee... the non-commissioned officer standing behind him will run him through with his bayonet and kill him on the spot.' The bayonet had not only become an offensive and defensive weapon, but it was now also being used to dispense summary justice in order to maintain discipline among the ranks even during battle. One of the new tactics developed by the Prussian Army of Frederick the Great was the 'Oblique Order'. Some believe that this tactic was one of the most important battlefield manoeuvres to be developed by the mid-eighteenth century. The execution of the tactic depended on rigid discipline, even under the most difficult of conditions. It was designed as an alternative to attacking the enemy head-on along the entire front. It was a shock tactic intended to catch the enemy off guard, and involved advancing in echelon against one of the enemy's flanks to outnumber it and overwhelm it before it could be reinforced. The way it worked was really quite simple and involved the Prussian troops advancing behind an advance guard, which kept the enemy's attention focused on them to the front. The following units were then given the signal to form 'oblique order' and move to one of the enemy's flanks and attack from the side. The manoeuvre was screened by the advance guard, which kept firing, and once the flank had started to give way under pressure the Prussian cavalry was sent in.

It was the 'blitzkrieg' of the eighteenth century and Frederick the Great was recognized as a master tactician, but this had not always been the case. It took time to formulate new tactics and train troops to carry them out, and at the Battle of Mollwitz on 10 April 1741 the Prussians won a hard-fought engagement against a force of equal numbers. During the Seven Years' War (1756–1763), despite the tactical advantages available to him with the oblique order, he is understood to have used it with success only twice. The first time was at the Battle of Leuthen in 1757, and again the following year at the Battle of Zorndorf. The Seven Years' War also involved France fighting against England and its Prussian allies. Prussian army drill was legendary in this period, and the main weapon of the infantry was the musket carried by each man and the bayonet fitted to the muzzle. At the time the Prussian army was given an advantage in firepower with the introduction of the iron ramrod. This replaced the wooden type, which could break in battle when reloading. It was also reversible, meaning that it could be used either way to reload the musket, rather than a man having to fumble with his ramrod to insert it into the barrel the correct way. This increased the rate of fire and other armies would soon follow the example. The Prussian army now instilled firepower and discipline as the elements to winning a battle. Those who witnessed the Prussian infantry being manoeuvred were left with a startling impression of the professionalism of the force. For example, during exercises in Silesia in 1785 a column of 23,000 Prussian infantry performed a wheel to a cannon-shot as the only signal for the manoeuvre to be made. Within seconds and without confusion, the column formed into a line some 2.25 miles in length.

Armies of the period used different formations when marching to battle. The British army, for example, marched in column but on reaching the actual battlefield the troops practised the tactic of deploying to form a line abreast to present a wall of musket fire. This tactic not only maximized the effectiveness of the firepower, but it also presented a 'hedge' of bayonets to counter an attack by cavalry, and also to meet head-on an attack by infantry. It was a tactic which would continue to be used until the late nineteenth century and was best known for its use at the Battle of Inkerman during the Crimean War. In fact, the manoeuvre would be immortalized in the painting called *The Thin Red Line* by the artist Robert Gibbs, which shows the 93rd (Highland) regiment with bayonets fixed receiving a charge by Russian cavalry. The British army developed the method where the troops could fire two or three

Private 5th West India Regiment with bayonet fitted to musket.

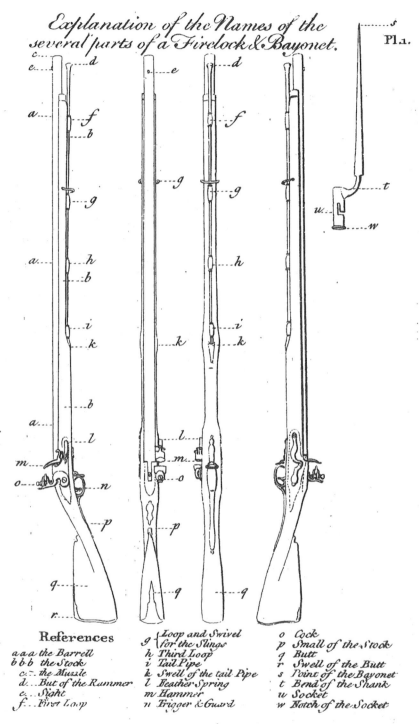

The British Brown Bess Musket with parts named and bayonet details.

volleys to 'soften up' the enemy and, having depleted the enemy's ranks, the British infantry would advance at point of bayonet, which became their stock in trade tactic on the battlefield. The French preferred to advance in column, which limited their ability to bring sufficient fire to bear against positions, and relied on the impetus of their advance to punch through. France and England would also come into conflict over the great wealth and resources of the vast nation of India. The British East India Company was established to conduct trade and created its own army to protect its interests from interference by France and Indian forces. At the Battle of Buxar on 22 October 1764, Hector Munro of Novar commanded around 7,000 troops to face an Indian force of 40,000 commanded by Shah Alam. During the course of the battle, Major Champion ordered the troops of his right wing to advance, but not to fire until they closed sufficiently with the enemy to allow them to 'push in' with the bayonet. The East India Company forces lost 1,847 killed and wounded, compared to 10,000 killed and wounded of Alam's forces and a further 6,000 were taken prisoner. It was an example of the steadfastness of better training and also an early example of the 'cult of the bayonet', and the action can be seen as being virtually a continuation of the 'push of the pike' from a hundred years earlier. Another advantage to deploying in line abreast against an enemy force which may be advancing in column was that it maximized the weight of firepower which could be brought to bear, and musket balls would tear into the leading ranks and either flanks of the column. With its narrow frontage, the advancing column could only manage a fraction of the firepower by way of answering the withering fusillade.

The eighteenth century was also to prove a period of experimentation for bayonet design, with armies across Europe trialling a range of different styles. Armies were also increasing in size and some countries, such as England and France, were sending troops overseas to protect the territories of their expanding empires, which would lead to fighting in the Caribbean and Far East. The flintlock was now the firearm of choice and two designs in particular would come to influence firearms more than any other types. These were the French 'Charleville' and the British 'Brown Bess', each of which would be produced in various forms, such as the 'long land pattern' for the Brown Bess. These weapons would be used in wars around the world and supplied to nations allied to either England or France. For example, during the American War of Independence, France supplied muskets to the Colonial troops fighting the forces of the regular British army.

Gunsmiths were now standardizing the external diameter of the musket barrel, each according to the national design, which meant that the weapons could accept the service-issue bayonet and the split socket design could be discarded. The new style had a solid tube forming the socket, which had a raised 'lip' called a 'bridge' that passed over the fixed stud on the barrel. All the infantryman had to do in order to fit his bayonet was to align the bridge of the socket with the stud, and the bayonet slipped onto the barrel. By giving it a twist, the zig-zag or mortise locked onto the stud and the bayonet was fixed. This method was a great improvement over earlier methods of attaching the bayonet, but it still remained far from being an entirely satisfactory method of attachment. During firing or retrieving the bayonet from the body of an enemy,

Recreated 9th Demi-Brigade French infantry armed with Charleville muskets fitted with bayonets.

the action could still make the bayonet work loose and it would become lost. Even when not in battle, the bayonet could still work loose or become difficult to attach as the metal contracted or expanded in the extremes of hot or cold weather. To overcome this problem, special locking rings and springs were often added at a later stage in production by craftsmen, after the weapon had been formed. Some armies introduced a range of 'locking' devices, which were extremely difficult to operate in cold, wet weather, such as the 'wing screw' device, which was used for a time by the Swedish army in 1698. The thread on the locking screw would have become stuck in cold weather and could also seize up with rust or dirt if not kept clean. In 1770 the French army adopted a locking ring, known as a 'basal' to secure the bayonets to muskets and the feature would remain in service until the spring-locking bayonet was introduced in the late nineteenth century. It was a simple feature which could be added to any socket bayonet and was copied by other countries including Spain, Italy, Prussia and the British army.

Recreated block of French Regiment de la Reine taking aim with their muskets which have bayonets fixed.

The shape and style of blades was also changing along with sizes. The British army had as standard a bayonet with a blade which was triangular in section. This style would remain in service for more than 150 years, but some regiments of the British army continued to use styles of bayonet which retained the traditional flat blade. There were also those types which were produced in very low numbers, and intended for use with specific types of firearms such as the style used on the Ferguson Rifle. At the beginning of June 1776, a young British army officer by the name of Patrick Ferguson serving in the 71st Highlanders demonstrated his newly invented weapon in front in front of several distinguished personages, including Lord Viscount Townsend, Lord Amherst and General Harvey at Woolwich in London. It was fired using the standard flintlock operation, but the innovation lay in its breech loading mechanism and the barrel, which had grooves known as rifling, cut in a spiralling form into the interior surface to increase accuracy. It used an improved version of the system developed by the French gunsmith La Chaumette, which employed a screw-in breech-plug which greatly speed up the loading process, and even allowed the firer to load the rifle whilst lying down.

Ferguson demonstrated his weapon successfully and it was decided to accept it into limited service. An order for 100 of the new rifles was placed, and Ferguson himself oversaw their production in Birmingham. Ferguson was also authorized to raise a group of men to train in the use of the new rifle and serve as sharpshooters. The weapon was also equipped with a bayonet and the rifle was used with good success during the American War of Independence. However, when Ferguson was killed at the Battle of Kings Mountain in South Carolina on 7 October 1780, the special unit was disbanded and the rifles withdrawn from service. A bayonet was developed for it, which was a standard socket-type of the time with the mortise cut into the sleeve. The blade was triangular in section with a 'fuller' groove and measured about 25 inches in length and 1.5 inches in width. It would appear from illustrations that it was an otherwise standard design in use with the British army. It is unlikely that the rifle would have become a standard service weapon, despite its high rate of fire and inherent accuracy, because of the amount of machining required in its production, which would have meant increased cost. The history of the Ferguson rifle is in itself interesting and serves as an example to demonstrate the fact that even specialist firearms with a small production run had to be equipped with a bayonet.

Around 1740 the French army briefly experimented with a type of socket bayonet, which had the blade mounted on a curved attachment to the sleeve, rather than being fitted with the more usual fitting with 90 degree angles. This was a short-lived trial, as was the bayonet with a blade formed into a half-tube shape to produce a gouge style. It was a socket bayonet in the true form and the blade would have produced a 'U-shaped' puncture wound. This was another design which was abandoned, presumably due to the difficulties in production. Bayonets in all armies were being made as a single piece, usually through a casting technique, but some were produced through the drop-forge method with the socket sleeve being 'braised' or welded on at a later stage in the manufacturing process. The best types of blades for these means of production were either flat blade or triangular section while anything more radical, such as the Durs Egg 'spear' type or the gouge blade, had to be made as individual items, which was time-consuming and more costly than production runs with thousands of numbers being turned out. The gouge bayonet may have been unconventional in appearance, but the blade was designed to produce a wound which was difficult to treat. A standard knife-type blade produces a clean wound with straight edges, which can be sutured or bound together tightly with a bandage, until the healing process begins. A rough wound such as that produced by a 'U-shaped' blade and even the triangular-section blade of the service bayonet was more difficult for surgeons to suture. Much later, the Russian army introduced a bayonet blade which was cruciform in shape and produced a cross-shaped puncture wound, which was equally difficult to treat with its many edges, to bring together to suture and close the wound.

In foundries and workshops all across Europe, thousands of bayonets were being produced on an industrial scale which was a world away from the artisanal method of making the original and earliest types of bayonets. Output had to be increased in order to meet demand to equip the armies fighting overseas, such as the British army fighting in Canada and the American War of Independence. Even in small garrisons such as those deployed for the defence of the island of Jersey in the Channel Islands, bayonets had to be made available. An inventory taken in January 1780 for the South-West Regiment records in the order book for the South Tower in the parish of St. Peter the items listed in its stores. This was one of the newly completed Martello Tower defences built on the island and included: '8 muskets, 8 bayonets, 12 cartridge boxes, [each] containing 48 cartridges and bullets. 8 musket flints, 1 screw driver, 1 sand clock, cleaning cloths for the arms. 1 bottle sweet oil for the arms, emery for the arms, 3X24-pounder cohorns or mortars. Pile of cannon balls for the cohorn.'

Small production runs of bayonets had to be paid for in order to cover the cost of materials and labour. For example, in 1789 the London-based gunmaker Henry Nock (1741-1804), devised a new type of bayonet with a locking ring and, according to sources, no more than 600 of these bayonets were probably ever made. Records show that the cost per item was two shillings and three pence (just over 11 pence modern pence) English money. The Nock bayonet still needed a locking ring, highlighting the fact that the problem of the bayonet working loose during firing had not been remedied. It was a problem which would bedevil all armies, and each in turn tried to find a solution in their own way throughout the nineteenth century. The French and British armies, for example, partly solved the problem by fitting a simple ring to the sleeve of the socket, and this could be rotated to 'lock' the bayonet onto the barrel stud. The Austrian army used a similar method, but the bayonet was fitted with a mortise slot, which ran diagonally across the sleeve, as opposed to the more common zig-zag pattern. The locking ring on the Austrian bayonet still functioned in the same way as other styles. By contrast the Danish army in 1794 employed a spring leaf device, called a Kyhl spring catch, developed by Johan Christian Wilken Kyhl, which was secured to the socket sleeve and clipped over the securing stud on the barrel. Trying to operate it in cold weather would have been a problem, but it must have proved serviceable because it remained in use for around fifty years. In 1830 Baron C.R. Berenger was experimenting with methods to prevent bayonets from becoming loose and even dropping off the barrel, and was granted Patent Number 5905 for his design. He believed that: 'the usual method of fixing bayonets they are inconvenient if tight, and if loose they are liable to fly off'. He devised a method whereby the base of the socket was fitted with a 'C-shaped' hook which engaged with the musket's ramrod. The shape of the ramrod being slightly tapered meant that in order to unfix the bayonet, the ramrod had to be withdrawn slightly. It seemed an unnecessary and fiddling method and the idea does not appear to have been introduced. The following year a certain Mr A. Demondion was granted Patent Number 6137 for his device for: 'A socket [which] is made to slide over the barrel, and when the bayonet is fixed the socket slides over the bayonet socket and is held in place by a spring'.

Troops in the army of the German state of Hanover were issued with muskets which had the bayonet locking device fitted a short way down the length of barrel from the muzzle. It operated in the same way as the Danish Kyhl's device, but in reverse and the method was also used in the Norwegian army at the time. Another method for securing the bayonet which originated in Hanover and was termed the 'Hanoverian Spring' was a short curved spring fitted to the fore-end of the musket barrel to grip the collar of the bayonet. After the Battle of Miani (also known as Meeanee) on 17 February 1843, soldiers of the British 22nd (Cheshire) Regiment recalled how Indian troops had managed to pull the bayonets from the British muskets. The battle, in modern day Pakistan, was fought between 2,800 troops of the East India Company commanded by Sir Charles Napier who faced a vastly superior force of 30,000 native troops commanded by Mir Nashir Khan Talpur. The tribesmen were fiercely brave but poorly armed with outdated weapons. The British were better armed and trained and this told in the battle, where discipline and firepower cost the Indian forces between 5,000 and 6,000 killed and wounded for the cost of around 256 killed and wounded. The fighting saw much use of the bayonet and some accounts tell how, even in their death throes having been stabbed by bayonets, Indian troops kept resisting. The troops had to use wire or twine to tie the bayonets on to secure them to prevent further such incidents. When news of the action reached London, it led to a Circular Memorandum being issued on 6 October 1844, stating that what was required was: 'A new and more secure method…for attaching or fixing the bayonet.' It was a problem which would be investigated many times between 1825 and 1853, when a satisfactory securing ring was at last developed.

The Prussian states would develop an almost universal 'bayonet mentality', with officers encouraging its use at every opportunity. During the campaign against the Austrians and Croats in 1756, during the Seven Years' War, the Prussian Duke of Bevern exhorted his troops with the words 'Shoot and get at them! Haven't you got bayonets? Go out and skewer the swine!' With modifications these various types of locking devices remained in use with European armies until the mid–nineteenth century. The cost of producing bayonets was economical compared to other bladed weapons such as swords, as shown in records from around 1815. An order for 5,000 swords to equip British troops armed with Baker rifles was submitted at a cost of 12 shillings and six pence (62.5 pence in modern British money). These were probably sword bayonets for the Rifle Regiment, which only 'fixed swords', as opposed to bayonets, when ordered to form squares against cavalry, or when preparing to receive an attack in closed formation. Another contractor in Birmingham at the same time supplied 5,000 bayonets of standard design at a cost of two shillings and eleven pence (about 15 pence in modern British money), and included locking rings and springs. This means that four bayonets could be supplied for the cost of each sword bayonet for the Baker rifle. The reason for the additional cost in the bayonets for the Baker rifles is because they were fitted with spring-loaded catches with which to attach them to the rifle. These were one of the earliest types to be made to this design and had to be assembled by hand, which meant more time to produce them, which in turn affected cost.

The 95th Regiment of Foot (later to become the Royal Green Jackets) was a specialist rifle regiment in the British army, and the hand–picked troops serving in this regiment were issued with the Baker rifle, which had a rifled barrel to give greater accuracy. It could be fitted with a special sword-bayonet with a brass hilt and knuckle bow (hand guard), which weighed 2lbs and measured 23 inches in length but some later versions were only 17 inches in length. The rifle was fitted with a bar to the right hand side of the barrel in order to allow the bayonet to be attached by sliding into the side of the brass hilt, so that it lay in the vertical plane. A spring catch locked it onto a notch on the attaching bar and it was very secure. Unfortunately, this placed it very close to a man's hand when reloading, and for that reason the rifle was very rarely fired with the bayonet fixed. These troops did not usually form part of the main advance in battle and so they had few reasons to fix bayonets, and tended to use them instead as a handy tool to chop wood or open packing cases. This was not without attendant dangers, as witnessed by Lieutenant John Kincaid of the 95th Rifle Regiment during the Waterloo campaign in June 1815, who observed two of his men using their sword bayonets to break up an old wagon for use as firewood. The wagon had been used to transport gunpowder and the bayonet of one of the men must have struck a metal fixture on the wagon and caused a spark, which ignited the remaining gunpowder, causing it to explode in a freak accident in which both men were killed.

However, when called on to do so they could fix bayonets and fight in close quarters as infantry of the line. The Baker rifle was considerably shorter than the Brown Bess musket and the great length of its bayonet, several inches longer than the standard socket type, meant that when fitted all weapons

Sword bayonet fitted to the Baker rifle.

Re-enactor depicting a rifleman showing the length of the bayonet, which the Royal Green Jackets Regiment still refer to today as swords, as fitted to the Baker rifle.

were of an equal length to defend against cavalry charges when formed into squares. Rifleman Edward Costello, who served with the 95th Rifle Regiment, believed that without the sword bayonet the troops of his regiment with their Baker rifles, which were slow to load due to the grooves of the rifling in the barrel, would have been helpless at close quarters against cavalry or infantry armed with bayonet and musket. The rifle troops armed with the Baker rifle were used in the role of skirmishers to scout ahead, and fired their accurate Baker rifles to shoot prominent figures such as officers. The accuracy of these troops was legendary and greatly admired, and tales of their achievements circulated among the troops. For example, Rifleman Thomas Plunket of the 95th at Cacabelos during the retreat of the British Army to Corunna in 1809 shot the French Général de Brigade Auguste-Marie-François Colbert at a range of some 660 yards, at a time when the best infantryman with a Brown Bess musket could hit a target at around 50 yards. The strange and unique regimental custom developed, and is still continued today in the Royal Green Jackets that the more usual order to fix bayonets is given as 'fix swords'.

The bayonet was now a standard issue weapon with troops and had already been proved useful in hundreds of battles across Europe. During the eighteenth

Sword bayonet as fitted to the Baker rifle.

The bar fitting on the Baker rifle for the sword bayonet.

century it would be used in many more actions and introduced to those armies which had never before encountered it. New drills and tactics would be devised to deal with the vast scale of new styles of warfare being conducted around the world, including India and North America where the use of the bayonet on the battlefield would prove decisive. One of those commanders to emerge and gain prominence across Western Europe was John Churchill who had been involved in helping put down the Monmouth Rebellion in 1685, and at the start of the eighteenth century would lead allied forces on campaign during the war of the Spanish Succession, but have come to be referred to by the more familiar term of the Marlborough Wars fought between 1704 and 1709. Later military commanders would emerge who had been influenced by his legacy of leadership, including officers such as John Burgoyne, James Wolfe and Colonel Robert Clive who, at the Battle of Plassey in India on 23 June 1757, would use the bayonet and better disciplined musketry to defeat an Indo-French force many times greater than his. Plassey and other battles of the Seven Years' Wars would help in finally adding the vast country of India to Britain's Empire. This was the time when the musketeer became a truly self-sufficient infantryman able to act and function on his own. All troops now had a role to perform on the battlefield, and they could be directed and committed to the fight in response to enemy movements. No longer were there large numbers of pikemen waiting for something to do; each infantryman was now armed with a musket. Commanders of all armies such as Maurice de Saxe realized this and the value of their troops meant they could fight at the best possible potential. These were the men who would dominate fighting in the eighteenth century, just as earlier commanders such as King Gustavus Adolphus of Sweden had dominated the seventeenth century, and he did not have the benefit of the bayonet.

Marlborough's Wars comprised of a series of campaigns, the stages of which are often referred to by the major battle which culminated from the manoeuvrings leading up to the engagement. Marlborough continued to move forward with the intention of engaging the Franco-Bavarian army which had crossed the river, and an action was fought at Donauworth which is sometimes known as Schellenberg. The battle commenced late in the day on 2 July 1704 and proved a hard-fought action. Just after 6pm Lieutenant-General Goor led a column of 6,000 infantry to attack the entrenched positions of the French. The English attacking force was formed up in three lines with eight battalions in support, eight more in reserve and thirty-five squadrons of cavalry. They approached uphill, which slowed their advance, and the French waited for them to appear. Commanding the positions was Jean Martin de la Colonie, who described what followed in his work *The Chronicles of an Old Campaigner*: 'So steep was the slope in front of us that as soon as almost as the enemy's column began its advance it was lost to view, and it came in sight again only two hundred paces from our entrenchments… The rapidity of their movements together with their loud yells, were truly alarming, and as soon as I heard them I ordered the drums to beat the "charge" so as to drown them with their noise, lest they should have a bad effect on our people.' This describes the obstacle facing the Allies, which would have exhausted the men as they climbed. Colonie continues: 'The English infantry led this attack with the greatest intrepidity, right up to our parapet, but there they were opposed with a courage at least equal to their own… It would be impossible to describe in words strong enough the details of the carnage that took place during this first attack, which lasted a good hour or more. We were all fighting hand to hand, hurling them back as they clutched at the parapet; men were slaying or tearing at the muzzles of guns and the bayonets which pierced their entrails; crushing under their feet their wounded comrades, and even gouging out their opponents' eyes with their nails, when the grip was so close that neither side could make use of their weapons. I verily believe that it would have been quite impossible to find a more terrible representation of Hell itself than was shown in the savagery of both sides on this occasion'

This is a first-hand account by someone who was in the thick of the battle describing close-quarter fighting, known as 'hand-to-hand' combat, and revealing all the drama of the struggle to repel an enemy using anything available, including bayonets and bare hands. The uphill attack was beaten back and the Allies lost many men killed and wounded in the assault. The bayonet was being used freely and, together with firepower from musketry and artillery, it had proven its worth in helping defend the French positions. In the end, though, the engagement was won by Marlborough but at great cost. Out of a force of 22,000 men Marlborough lost 3,741 killed and wounded representing 17 per cent of his force. The French and Bavarians had a combined force of 13,000 men and at the end of the fighting they had lost 5,000 killed and wounded, representing just under 40 per cent of their army and a further 3,000 were taken prisoner. At the Battle of Malplaquet fought on 11 September 1709, the casualty figures were much higher. Marlborough commanded a force of 86,000 men and lost 21,000 killed and wounded, which represented more than 24 per cent of his force. The French and Bavarians fielded an army of 75,000 men and suffered 11,000 killed and wounded, which represented less than 15 per cent of their force. An examination of the French and Bavarian casualties revealed that around 66 per cent were caused by musket fire, and of this figure 60 per cent had been hit in the left hand side of their bodies, indicating they were in the act of firing towards the British lines. By contrast, only 2 per cent of the wounded were recorded as having been caused by bayonets, which represents around 220 men. The number of those killed by bayonets would have been among those 34 per cent killed by other causes such as artillery. The type of victory won by Marlborough at Malplaquet was not the kind of thing the British or their allies could afford to repeat.

As armies increased in size and more infantrymen were armed with muskets, this led to the infantryman's independence being reaffirmed and the strength of battalions were now being described in terms of the numbers of bayonets they could deploy. By looking at the numbers deployed for battle, it is possible to see how influential the bayonet had become. Seeing an army manoeuvring on the battlefield with bayonets fixed would have been impressive and intimidating in equal measure. To the uninitiated recruits going into battle for the first time, such a sight would have been terrifying as they would have imagined the many ways in which they could be wounded. This was where discipline told and the ranks of the better drilled infantry would hold firm, despite being bombarded by artillery, and remain in their positions to meet advancing lines of infantry who would also have their bayonets fixed. As the two sides clashed and crossed bayonets, such an encounter was the defining moment of the engagement. The weaker and less disciplined side would invariably break ranks and retreat, whilst the more determined side would clear the positions and hold the ground.

The bayonet was recognized first and foremost as being a weapon designed to inflict wounds or kill by means of producing puncture wounds through the action of stabbing. It was and still remains a close-quarter weapon and, although originally intended as a defensive measure, a series of drills were developed which helped turn it into an offensive weapon, also. One of the military commanders responsible for helping bring about such a change was Lieutenant Colonel James Wolfe, who was a serving officer in the British army, and it was under his personal direction that new tactics were introduced which changed the role of the bayonet and led to it becoming an offensive weapon to be used in a massed charge when engaging the enemy to push them off the battlefield. In 1750 at the age of just 23 years old, James Wolfe was promoted Lieutenant Colonel of 20th Regiment of Foot. He was the protégé of the Duke of Cumberland and, despite his very young age, he had already seen considerable service on campaign. Born on 2 January 1727 in Kent, James Wolfe was the son of a Lieutenant General. Despite his youth, he enlisted in the 1st Marines Regiment, his father's regiment, in 1738. He did not have to wait long for his first experience in battle, which came in 1742 during the War of the Austrian Succession, which had broken out in 1740. Among the British troops sent to support Austria was the 1st Marine Regiment in which Wolfe was serving as a Lieutenant. The war would last until 1748, but Wolfe transferred to the 12th regiment of Foot (later to become the Suffolk Regiment), and he campaigned with it in Flanders. He saw much action and was thrown to the ground when his horse was killed at the Battle of Dettingen on 27 June 1743. It was during this battle that he first came to the notice of the Duke of Cumberland, and a year later Wolfe was made a Captain in the 45th Regiment of Foot (later to become the Sherwood Foresters).

In October 1745 Wolfe's regiment was recalled to England and sent to Scotland as part of Cumberland's force to help suppress another Jacobite rebellion, and Wolfe was present at the Battles of Falkirk on 17 January and Culloden on 16 April in 1746. He would almost certainly have seen troops using the bayonet in earlier battles, but at the Battle of Culloden we know from his later writings that he certainly witnessed first-hand it being used with the utmost effectiveness against the brave but ineffectually led Scottish Highlanders, as they attacked the better trained and disciplined troops of the British army. At Culloden British troops directed the thrusts of their bayonets to the right-hand side of their Scottish Highlander opponents; in other words, the man's sword arm, because that was the side which was unprotected, unlike his left side, which was covered by a large shield. The optimum moment to thrust, given the opportunity, was just as the Highlanders lifted their sword arms to strike and the bayonet could be directed to pierce under the armpit, which could puncture the lung and possibly either the axillary or subclavian arteries – wounds to any of which would certainly prove fatal. Of course, the troops did not have to be instructed in the subtleties of the biology of the human body, and would have simply been trained and encouraged to treat it as any other target to be attacked.

The instructions for bayonet drill taught to Cumberland's troops included the direction that a man with bayonet fitted should: 'Step forward about eighteen inches with the left foot, bending at the left knee, and at the same time seizing the butt with the right hand (placing the plate full in the palm of the hand) bring down the muzzle so as the firelock may rest upon the left arm, almost level, and as high as your breast, the left elbow turned out towards the front, the fingers and thumb towards the lock.' If necessary, the next order of command would be to 'push bayonets' to engage an advancing enemy. This position with the supporting elbow at 90 degrees would remain in use throughout the eighteenth century and is shown in illustrations of the time, such as a sketch made by a Lieutenant Baillie and a painting by the artist David

34th Regiment of Foot circa 1740. Later to become Lucas's Regiment and then the Border Regiment. He is carrying a bayonet on his waist belt.

Morier. We can also see how the movement may have looked for real in the 2003 movie *Pirates of the Caribbean: Curse of the Black Pearl*, starring Johnny Depp. There is a scene in the film where Johnny Depp's character, Captain Jack Sparrow, is taken prisoner at point of bayonet by British troops armed with Brown Bess muskets. The troops are shown wielding their muskets and bayonets in the manner as laid out in Cumberland's instructions. The musket held in this way can be moved up and down and pivoted left and right with minimum of effort, so the infantryman can defend himself against attackers from all directions. It also means the technical advisers on the film certainly knew the tactics of the infantryman of the eighteenth century, which is more than can be said for Morier's painting depicting the use of the bayonet at Culloden in a work entitled *An Incident in the Rebellion of 1745*. This painting shows an infantryman using his musket fitted with a bayonet in an overhand downward thrust, which was unusual, and with both arms raised his torso is left exposed and vulnerable to a sword thrust. Morier enjoyed the patronage of the Duke of Cumberland and joined him on campaign in Scotland. It is not known if he was present at Culloden or even witnessed any actual fighting. His depictions would indicate that he did not, because the artist's imagery shows he was unfamiliar with his subject and indeed may never have seen a soldier in battle and was using more than a little artistic licence.

The peace of Britain became threatened in July 1745 when Charles Edward Stuart landed in Scotland with a retinue determined to raise an army and take up his father's old cause from 1715, which was to fight for his claim to the throne. Supporters came in from the Highland clans and he soon had a force of around 1,500 troops. This was the beginning of the second Jacobite Rebellion. The British Government responded to the threat and sent a force of 4,000 troops under the command of Sir John Cope with orders to supress the rebellion. Cope marched from Stirling and finally met the Highlanders who were in well-prepared positions at Corrieyairack. He decided against attacking and instead decided to march to Inverness, where he put his troops on ships and sailed to the Firth of Forth. From here he marched inland and eventually encountered the Jacobite force at Prestonpans on 21 September 1745. During the ensuing battle Cope was soundly beaten giving Charles' army a great boost to morale. By November his army numbered around 5,500 and he marched south into England and reached the city of Derby on 4 December.

Unfixing bayonet and returning it to the scabbard.

Volunteer firing his musket. Troops were trained to shoot with bayonets fixed.

There was no credible force to intercept Charles' Jacobite army of Highlanders and various mercenaries, but suddenly on 6 December he decided to withdraw and move back to Scotland. There have been various reasons put forward to explain his decision, with the most obvious being that he realized he could expect little, if any, support for his cause from the English. This unexpected move gave the British army time to move troops against the threat and the Duke of Cumberland led his troops to pursue the Jacobites back into Scotland. His advance guard harried the rebels and a skirmish ensued at Penrith on the 18 December. On the same day at Clifton Moor, Highlanders charged a small group of British soldiers and sent them fleeing. A month later on 17 January 1746 the Highlanders encountered British troops at Falkirk and the weather was so bad they could not load or fire their muskets. The Highland charge was put in with such force that the British fled the field again and reaffirmed the belief in the headlong charge with swords.

Meanwhile, Cumberland was laying his plans and training his troops so that by April 1746 he was ready to make his move. The British troops were well trained and disciplined and were also being trained in a new tactic using the bayonet by lunging to the right, which was intended to catch the Highlanders at a disadvantage as they charged. Normally, an infantryman would engage the man directly to his front and lunge at him; that would turn into single combat between a man armed with a sword and the other armed with a bayonet and musket matching one another with thrusts and counter-thrusts. By thrusting to the right it was hoped that such an unexpected move would catch Highlanders unprepared for such a movement and against which they had no defence. The Highlanders had proved themselves to be formidable enemies and it was known from previous battles that each man could carry a number of weapons including pistols, daggers, 'claidheamh mòr' (to mean 'great sword' and known as the claymore) and the smaller broadsword, which was the main weapon for close-quarter fighting. In their left hands men carried the 'targe' or shield to protect them against swords, or bayonet thrusts. The Highland charge could break through ranks of infantry using the elements of weight and speed along with the ferocity of the men. The Highlanders had developed a simple yet effective way of using their shield to defeat a bayonet thrust. As a man approached, he would drop to one knee and as his opponent lunged at him with the bayonet, the Highlander would lift his shield to deflect the thrust upwards. It was an instantaneous action which had to be completed

Changing sentry in camp, circa 1780.

without hesitation and, if done properly, the movement was so unexpected that it would leave the enemy's torso exposed, allowing the Highlander to stab it with his sword.

On 8 April Cumberland began to march north from Aberdeen, and on 16 April he faced the Jacobite army of Charles Stuart across the battlefield at Culloden. Cumberland had 8,000 well-trained and disciplined troops supported with artillery. Charles had 7,000 Highlanders and no artillery and they relied on their sudden charge to bring them yet another victory. The British troops had also been instructed to hold their fire until the Scots were only ten yards (30 feet) range before firing, and at such close quarters they were sure to hit their target. The artillery also fired using 'grape shot', which comprised of several or more iron balls about the size of an egg held together with either wire or hessian material which, on being fired from the cannon, would spread out. The effect was devastating at close range and after one hour of fighting the battle was over. The English had lost 50 men killed and further 259 wounded. The Scottish Highlanders lost between 1,500 and 2,000 killed and wounded with musket fire, artillery and bayonet at close quarters.

Charles Stuart fled the battlefield and the Jacobite Rebellion was over, with the bayonet contributing a part in securing the northern borders between England and Scotland. Many lessons had been learned and as news of the war and the accounts of the battles spread across Europe, armies throughout the continent developed or changed tactics to make better use of the bayonet. They also developed ways of dealing with bayonet charges. At this time across the length and breadth of Europe, from Spain in the west to Russia in the east, battles were being fought as part of various campaigns or wars by armies of ever-increasing size – all of which meant more bayonets had to be produced. Armies of 80,000 men and even larger were not uncommon and when two sides of similar strength met, it meant hundreds of thousands of bayonets would be on the battlefield all at the same time and contributing to the casualty lists.

The effectiveness of the bayonet in battle would always be questioned, and never more so than during the last battle of the Jacobite Rebellion. Some sources have now begun to doubt the effectiveness of the new bayonet drill used in the Battle of Culloden, and look closely at the numbers of casualties inflected by the bayonet at the time of the battle. Some views now express that the Highlanders' charge did not impact on the British ranks all at the same time, but rather in a wave-like effect along the front rank as the slower men made contact after the faster men, like runners in a race. This has led to debates concerning the effectiveness of the drill to stab to the right. There is something in the opinion of doubt, but the fact remains the British lines managed to hold even though the Highlanders did penetrate the ranks in some places. Musket fire almost certainly inflicted the greatest number of casualties along with the artillery, but the bayonet helped the ranks hold their positions and the weapon's reputation was enhanced along with that of the British troops, who remained positive during the battle. British troops would have found it strange to thrust to the right instead of straight at the target directly to a man's front, but soldiers in all armies become used to new drills through instruction. After the battle Wolfe wrote his opinion of the bayonet during the engagement, stating that: 'Twas for some time a dispute between swords and the bayonets; but the bayonet was found by far the most destructible weapon' His mind was made up when he saw how the bayonet could be used as a weapon which could sway the course of a battle, provided, of course, that the troops were reliable men.

In January 1747 Wolfe, now aged 20 years, was once more engaged in the War of the Austrian Succession and over the next six months he took part in further engagements such as the Battle of Lauffeldt on 2 July 1747, where he was wounded. He returned to England and engaged himself in writing military tracts and pamphlets on training. He had served on six campaigns and fought in four pitched battles and he was only nine years into his military career. He was full of innovative military ideas and in 1755, was in his fifth year holding the rank of Lieutenant-Colonel and in command of the 20th Regiment of Foot (later to become the Lancashire Fusiliers). It was the perfect position, which allowed him the opportunity to emphasize his own opinions concerning the importance of musketry with target practice to improve accuracy, along with bayonet drill when at close quarters with the enemy. Three years later he was again on campaign, this time in Canada where British and French forces were fighting over the vast territories and the wealth it was thought to contain. It was a strange war, with European armies fighting battles using European tactics and with local natives acting as scouts. During the fighting, soldiers would engage in combat with local natives using bayonets against primitive clubs.

In June 1758 Wolfe was present at the siege of Louisbourg, during which action the British facing French regular troops were ordered to 'march up close to them' then they would fire their muskets at very close range, after

General Wolfe, who developed new bayonet drills for battle.

Light infantryman with musket and bayonet fitted at practice drill.

which they were to 'rush upon them with their bayonets.' The French were more than capable of giving as good with the bayonet as the British were, as demonstrated when they charged a position in the siege lines around the town and managed to seize the redoubt when the British were 'forced on by fixed bayonets.' The British were not going to sit idly by and allow this to happen, so a counter-attack was ordered. The position was recaptured by the 22nd Regiment of Foot (later to become the Cheshire Regiment) without a shot being fired, as recalled in a letter to his wife by Major Alexander Murray, telling her how the troops 'with their Bayonets, drove them [the French] out of the work without a shot till they began to run'. After the Louisbourg Campaign Wolfe returned to England on sick leave, but returned to Canada in February 1759. In June that year the British moved forces towards the French-held city of Quebec on the St Lawrence River and Wolfe, by now a Major-General, was in command of the force.

Wolfe made his preparations to attack the city and skirmishes were being fought across the Plains of Abraham where the city stood. In July one of these actions involved a British unit called the Louisbourg Grenadiers, which had been raised by Wolfe himself. They charged after the French troops to complete their rout from the battlefield. After the fight Major Alexander Murray, serving with the 45th Regiment of Foot, wrote home to his wife telling her in his account how he had seen 'bayonets were bent, and their muzzles dipped in gore.' Fighting continued elsewhere, such as the skirmish on 24 July 1759 at La Belle Famille, when about 130 men of the British 46th Regiment of Foot commanded by Lieutenant-Colonel Eyre Massy engaged around 800 French regular troops. He later recalled how his men received the enemy with 'vast resolution', and never fired a shot until they could reach them with the bayonet. All the while Wolfe was getting ready to attack the city and finally on 13 September he personally led his men and scrambled up the rocky face to reach the Plains of Abraham, where he assembled his force of around 5,000 men. During the fighting Wolfe and his French counterpart, the Marquis de Montcalm, were both fatally wounded. The city surrendered on 17 September and British control in North America was assured, for the time being at least, and had been achieved partly through the use of the bayonet.

The British did not always have it their own way, though, when it came to bayonet fighting as observed by James Johnson, a Jacobite living in exile and serving as an aide-de-camp to the French Chevalier de Levis during the fighting in Canada, and who witnessed French troops charging the British from their positions at Sainte-Foy on 28 April 1760. In his journal, John Knox who served with the British forces and participated in the battle, throws doubt on the efficacy of the French use of bayonets during this engagement and wrote: 'I have frequently had the honour of meeting them [the French] in the course of my service, and I never saw them disposed to come to the distance of a pistol shot, much less bayonet pushing'. This is another personal observation and contrasts with others who claim that it was the bayonet which made all the difference. Such conflicting opinions are only to be expected when it comes to assessing a weapon as controversial as the bayonet. For example, Wolfe believed implicitly in the use of the bayonet and yet here is a soldier claiming that its use did not make an impact on the outcome of a battle. Of course, this is only one man's opinion on a single engagement and the use of the bayonet has to be considered much wider and cannot always be given credit.

At this time the bayonet fitted to the musket was still being used in virtually the same way as pikes had been 100 years earlier. The Prussians had already developed a set of bayonet drills for use on the battlefield and although new drills had been used at Culloden, it would take commanders such as Wolfe to set new standards by emulating existing proven bayonet drills and perhaps even improving on them. Imitation is the sincerest form of flattery, so it is said, and by copying the Prussian system, Wolfe had seen to it that the best bayonet drills were introduced to the ranks of his troops, and this was a legacy which would be passed on to other commanders. The new bayonet drill being taught to new recruits in the Prussian system involved holding the musket in both hands at waist height. It was an ideal method because as the line advanced, it presented a row of bayonets all at the same height and in line with the torsos of the enemy. This meant that they could thrust forward into the abdominal area and cause deep penetrating wounds to vital organs such as the liver and even rupture the stomach. The stance also meant the musket and bayonet was presented at the optimum height to parry a thrust from an enemy's bayonet and still allow the line to advance, and by sheer weight of numbers and strong discipline, force the enemy away from their positions. It was a tactic which was later to be copied by armies across Europe, and was even illustrated in the *Treatise* by William Wyndham around 1760. At this time the bayonet with its long, tapering blade design, triangular in section, was beginning to evolve and gain wider acceptance. With its 'fullered' grooves it was made lighter but still retained its strength and rigidity. The design would remain a favoured pattern and continue in service until the late 1880s, having served all the way through the Napoleonic Wars, Crimean War, American Civil War and many other later conflicts.

As France and England emerged as the two dominant powers of the European states with designs to create empires, their interests in particular would bring them into conflict. They would also, each in turn, have the local populace to deal with. For the most part, these were poorly armed civilians of native forces with little or no formal military training, and what they lacked in leadership skills they made up for with determination. Unfortunately,

Sentry of 1st Foot Guards on sentry duty with fixed bayonet, circa 1760.

this was not sufficient to withstand a well-trained army with modern weapons. One area where France and England had troubles was in the Caribbean islands where slaves were used on sugar plantations and revolt was a threat, such as St Kitts in 1639 and St John in 1733. In 1762 England found itself engaged in a campaign against Spain on the Island of Cuba during an extension of the Seven Years' War. It was a bloody affair and involved fierce fighting, during which bayonets were used. Indeed, first-hand accounts of bayonet fighting are extremely rare, but from this campaign we have been left with such a document.

Bayonets fitted to muskets are ill-suited to fighting in confined spaces due to the length of the weapon, which can be as much as six feet. But when checking to see if a building has been cleared of an enemy, it is sometimes the case that a soldier has to enter a house and it can lead to him being surprised. It can also be the case where an enemy is pursued into a building, as was the case when British troops were fighting on Cuba in 1762. Troops of the 2/42nd Regiment were confronted near some buildings just outside the island's capital, Havana. John Grant serving with the regiment later recalled how he was chased into a building by a man but was able to fire at him, which instantly filled the interior with thick smoke. The shot had hit his pursuer but, being a trained soldier, Grant did not let his guard drop and prepared to defend himself at point of bayonet and he lunged forward. He later wrote how: 'an agonised groan soon informed me that my aim had been too fatal, I found my fuzee [musket] grasped, when attempting to withdraw it, and the smoke clearing, I found I had run my opponent through the body and fairly pinned him to the wall, whilst his face glaring with rage and pain, was close to me.' Close-quarter fighting does not come more personal than this account with one-on-one inside a house. Grant would have been terrified but, being a soldier, he knew how to defend himself. It would have been an unforgettable experience, no matter how many more fights he was involved in.

At the beginning of the eighteenth century the tactic of firing muskets in volleys to maximize effect was still in the process of being developed, but once it had been tried and proven to be effective the method gained wider acceptance. The long-held practice in many armies was to open fire and then each man would load and fire at his own speed. This reduced the overall effectiveness and left the musketeers open to attack from cavalry, and with no pike blocks now to protect them, the musketeers had to use their bayonets and this led to the development of the method of firing en masse. For example, the American War of Independence was not yet a year into the fighting when the British officer General John Burgoyne issued in a General Order a proclamation on 20 June 1777 in which he pointed out that: 'The officers will take all proper opportunities to inculcate in the men's minds a reliance on the bayonet; men of their bodily strength and even a coward may be their match in firing. But the bayonet in the hands of the valiant is irresistible.' Indeed, this was a commander who very firmly believed that the success of the British army rested greatly on the use of the bayonet, but firepower was also a deciding factor.

When this war started in 1775 the British army had been using bayonets for almost 100 years and all troops who served in the war were issued with them, which the American colonial forces found out to their cost. Despite this recognition of how useful the bayonet was on the battlefield, especially for clearing away any remaining troops, the bayonet was not standard equipment in the Continental Army. Indeed, it would not be until the mid-1840s that American service rifles would be issued with bayonets for the Model 1841 rifle. This deficiency has been seen as a weakness during the War of Independence when troops of the Continental army under General George

Washington found themselves enjoined by the British troops, who advanced across the battlefield with fixed bayonets to drive them off and hasten their retreat. During the War of Independence, Captain Hind wrote a book called *The Discipline of the Light Horse*, published in 1778, and in it he describes how dragoons were armed with carbine and bayonet.

General John Burgoyne, born in 1722, would become one of the leading field commanders of the war and was an experienced field commander, having seen service during the Seven Years' War and fought on campaign in Portugal during 1762. In America he faced French forces, which had been sent to support the American colonial forces, and he observed how at one encounter: 'The French came up the hill with a brisk and regular step, and their drums beating pas de charge: our men fired wildly and at random among them; the French never returned a shot, but continued their steady advance. The English fired again but still without return.' He continued 'and when the French were close upon them they wavered and gave way.' Another British officer describing the action in a graphic manner wrote how: 'The French regiment formed close column with grenadiers in front and closed the battalions'. He continued his description of what is almost certainly the same engagement as recounted by General Burgoyne. 'They then advanced up the hill in the most beautiful order without firing a shot... When at 30 paces distant our men began to waver, being still firing... The ensigns advanced two paces in front and planted the colours on the edge of the hill and officers stepped out to encourage the men to meet them. They [The British troops] stopped with an apparent determination to stand firm, the enemy continued to advance at a steady pace and when quite close the Fusiliers gave way: The French followed down the hill on our side.' The French troops would have presumably had fixed bayonets as was the preferred method at the time, and with these they were able to achieve the rout and drive the British troops from their positions. Despite the discipline in their ranks, the British troops knew they were beaten and at point of bayonet were driven away from their positions. It was the classic tactic which had been used many times before, and would be used again many times in the future.

The socket bayonet was now universally being fitted with a 'Z-shaped' slot, sometimes referred to as a 'mortise' cut, into the sleeve, which slid over the end of the barrel and securing rings fitted to prevent the bayonet from sliding off. The blade was 'cranked' to the side, which meant it was offset from the barrel by an angle of 90 degrees, giving sufficient room to allow the infantryman to reload without catching his hand when using the ramrod, and to all intents and purposes the bayonet looked perfectly developed. However, there was still room for improvement and some designers thought they could enhance its usefulness. For example, the 'bend' by the blade was known as the elbow and on some designs, such as the English style, there was a small disc-like protuberance called the 'shoulder', beyond which was the bayonet blade proper. This otherwise inconsequential part of bayonet was considered as having no possible use, but in 1816 Mr F. Richardson was granted the Patent Number 4031 to develop a small serrated edge which fitted into this bend with the intention that it be used to cut off the end of the paper cartridges during loading. It seemed like a good idea, at least to Richardson, but the troops preferred to use the tried and much easier way, which was to simply bite off the end of the paper cartridge, and this was also a much faster method. Richardson's idea became no more than a novelty and was never adopted because there was no requirement for it when teeth were readily available.

The bayonet blade was developed into triangular form in section from around 1715 and the length varied between armies. This basic design probably originated in France also and proved to be stronger than the traditional flat-bladed knife-like design. The benefit of the new triangular section was recognized and armies across Europe began to adapt it, including the British army. The blades of the new design bayonets were increased to measure between 15 inches and 18 inches in length or approximately half again as long as the first designs. Those few extra inches may not seem important, but the advantage they gave on the battlefield would provide the infantry with an advantage by reaching out just that little bit further. Like a boxer with a longer reach than his opponent, the longer blades allowed the bayonet to be driven into the target before the enemy could make contact. This was important against cavalry, because if the infantryman could get under the rider's sword he was then in a position to stab him with the bayonet. If the rider could not be stabbed then the horse, being the larger and more vulnerable target, would be attacked. Once the horse was down the rider was also on the ground, where he could be dealt with either by use of bayonet or axe, which some infantrymen carried.

Despite the usefulness of the bayonet, not all armies used it as a standard item for troops. In 1775 Britain faced an enemy comprised mainly of civilians who were not trained in war. This was the war in the American colonies and before the year was out, British troops had fought troops of the American

Continental Army at engagements such as Lexington and Concord, where they crossed bayonets. At the Battle of Bunker Hill on 17 June the Colonials threw rocks at the lines of advancing British troops when they ran out of ammunition. Congress appointed General George Washington to command the Continental Army and he was able to use his knowledge of British military skills, having served in the British army and fought on campaign during the Seven Years' War. Within a year, skirmishes had turned to pitched battles and the Continental Army, despite having instruction in fighting the British from experienced European officers, suffered terrible losses on the battlefield. The regular troops of the Continental Army were issued with bayonets whenever they were available, but those serving with the militia were not issued with the weapon as standard. One of those foreign officers giving military instruction to the Continental Army was General Friedrich Wilhelm von Steuben, who had served in the Prussian army and campaigned during the Seven Years' War. He instigated a stringent training programme for the colonials, instructing them in the use of the bayonet and also to improve their musketry. He was at Valley Forge, which served as the main camp for the Continental Army during the winter of 1777–1778, at a time when the morale of Washington's was at its lowest ebb, but they had not given up the fight and he observed that their muskets were in a: 'horrible condition, covered with rust, half of them without bayonets.' He knew that if the Continental army was to continue the war and fight, it had to be taught European standards and he ordered that each man had to be issued with a musket and bayonet if they were to meet the British on anything approaching equal terms.

Production of weapons by the Colonists was virtually non-existent and muskets had to be purchased. Later, the French would supply weapons to the Continental Army along with other stores and troops. Weapons and equipment were also captured from British garrisons such as the stocks seized following the attack against the barracks at Trenton on 26 December 1776, where a force of 1,500 Hessian troops fighting for the British was garrisoned under the command of Colonel Johann Rall, who exhibited his ignorance of the danger of attack by dismissing the need to dig trenches to defend the location by announcing: 'Let them come! We want no trenches; we'll have at them with bayonets!' In a daring operation Washington's men rowed crossed the Delaware River in small boats during the early hours of the morning, taking advantage of the poor light and the fact that the Hessian troops were still hungover from the previous day's drinking bout to celebrate Christmas. The Colonial troops took the garrison by surprise, but even so the fighting lasted an hour, during which Colonel Rall was mortally wounded and about 1,000 of the garrison fled leaving the rest either dead, wounded or to be taken prisoner. The action was fruitful and yielded up 1,000 muskets and bayonets along with other supplies, including gunpowder, 40 horses and six pieces of artillery.

It was probably due to lack of manpower that prevented the Colonial forces from being drawn into mounting bayonet charges, which would have produced high numbers of casualties as they advanced across the open ground of the battlefield. The British army had plenty of reserves to replace those troops and weapons lost in battle, whereas the Colonial forces, being made up of volunteers, found it difficult to replace the fighting men. The regular troops of the Continental Army received basic training due to the efforts of General von Steuben, and this now included bayonet practice ready for when they came into close contact with the British troops. The British believed this refusal to cross bayonets was because the Colonial troops, not being regular troops in the true sense of their terms, could not or would not face up to the British troops in their red tunics as they marched forward with bayonets fixed. The instruction in the use of the bayonet under General von Steuben's direction paid off during the night attack made by a force of 1,500 Colonial troops at the Battle of Stony Point on the 16 June 1779, against a British force of 750 men. The attackers went into action with unloaded muskets to prevent any accidental firing and the positions were carried at the point of the bayonet. In overall terms the action was a rather minor affair in the war, but it showed what could be achieved by use of the bayonet alone. The Colonials killed 20 men, and captured 472 unwounded and a further 74 wounded. They had in turn lost 15 men killed and 83 wounded. All the British casualties were produced by bayonets and other hand-held weapons, including axes.

Two years later, any belief that the Continental Army did not want to use bayonets was further shattered at the Battle of Guildford Court House on 15 March 1781. During the attack the British 23rd Regiment of Foot hesitated during their advance when confronted by Colonial troops levelling their muskets at them from behind a wooden palisade. Colonel Webster urged the men on and they charged forward and a bayonet fight at close quarters ensued. A famous painting of the engagement by the artist H. Charles McBarron shows American troops holding their muskets fitted with bayonets facing the British advance. The 5th Maryland Regiment was not experienced in battle and the troops broke when attacked by the British. They had to be supported by William Washington's Dragoons, but the 1st Maryland Regiment were more experienced and not only held their positions, but counter-attacked with a bayonet charge, which threw back the British troops. Some paintings by other artists depicting the same action show American troops with muskets fitted with bayonets, and others depict the troops without bayonets fitted to their muskets. This may seem like an anomaly, but in fact both depictions are correct. Regular troops in the Continental army were not always equipped with bayonets and stocks would have been added to when supplies were captured from British stocks, as happened at Trenton. The Continental army would stand and fight on the battlefield and paintings showing them with bayonets fixed are correct. Troops shown without bayonets may not have been issued with them or they may have been lost. Colonial troops depicted fighting guerrilla tactics in the dense woodlands are not shown using muskets with bayonets; this is because it was known that the extra length added to the already long barrel would have been an unnecessary encumbrance among the trees. Washington and his officers were experienced enough to know when it was best to draw up their troops and fight a battle using European-style tactics if they had sufficient strength, and at other times when they had neither the troops nor the supplies, it was best to avoid engaging in such an action. Set-piece battles such as Bunker Hill, Guildford Court House, Concord and Lexington could not be avoided, and they had proved to be costly affairs in the numbers of troops killed and wounded.

The Colonial militiamen who lived in the wilderness, known as frontiersmen, used a type of flintlock rifle produced on a design known as the Kentucky long rifle. It was a civilian weapon designed for hunting, and the longer barrel gave it increased range and better accuracy over the standard issue Brown Bess musket used by British troops. These frontiersmen cast their own bullets because the calibre of the weapons could vary according to the manufacturer, and also resupply of ammunition was not possible in the remote regions. They often made the lead balls for their muskets slightly smaller than the calibre of the barrel and made up the difference by wrapping the lead ball in either a piece of greased leather or linen patch to grip the rifling and impart spin to the ball to give it stability when fired, and this improved accuracy. These muskets not being of a regulation size did not lend themselves to being fitted with a bayonet and, as long-range marksmen, they were never intentionally going to 'cross muzzle or blades' with British troops at close quarters. Plug bayonets may have been an option, but such irregular troops would have wanted to keep their weapons in a state of readiness to fire at a moment's notice. They did not wish to clog the barrel with a bayonet for which they would have only had a use for as a tool for chopping wood or as a cooking implement.

The night attack at Stony Point was a success, but it was not the first time such an action had taken place and indeed was probably inspired by similar actions which had occurred two years previously. The war in America was unlike anything before encountered and some British officers went against convention and devised cunning surprise attacks to catch the Colonial forces off guard, and even went to great lengths to plan such operations, which also included night attacks on encampments. During such attacks, the officers would order their men to advance with unloaded weapons but with bayonets fixed. Major-General Charles Grey, later 1st Earl Grey, took this one step further and ordered his men to remove the flints from their muskets before a night attack on an American encampment at Paoli Tavern on 20 September 1777, where a force of 1,500 regulars and 1,000 militia were encamped. Grey's 1,200 troops attacked mainly using the bayonet, which produced horrific wounds during the close-quarter fighting, and which later led to the engagement being called a 'massacre'. Some would have used axes, which were favoured in such close-quarter fighting, and at the end of the action 53 Colonials were killed, 113 wounded and 71 taken prisoner. They had been unable to defend themselves properly during the time taken to reload their muskets because many apparently lacked bayonets, which could have been used at close quarters. The British lost four men killed and seven wounded in the fight. Some of the Colonial dead were described as 'mangled', to which charge it was explained that the bayonet is a messy weapon. Following the action, Major-General Grey earned for himself the nickname 'No-flint Grey', a term which was emphasized a year later when he ordered his men once again to remove the flint from their muskets when they attacked an encampment of dragoons at Baylor on 27 September 1778. At point of bayonet some sentries were eliminated and the remainder of the unit captured at point of bayonet, and a number were stabbed with bayonet thrusts. Because of the methods used, the engagement was also called a 'massacre' by the Colonials.

The bayonet was used in many regular skirmishes and battles during the course of the war, such as at Brandywine Creek, for example, where on 11 September 1777 some British regiments used only the bayonet to attack an

American unit whilst it was asleep, and managed to kill a number of Colonial troops without disturbing the camp. The war lasted for over eight years and, despite Britain's best effort, it was lost and the financial costs were enormous, as was the casualty list. It was a war in which strident changes were made in weaponry and battlefield tactics. The most basic of all the weapons was the bayonet, which as an otherwise conventionally designed bladed weapon was in keeping with military tradition of using such weapons like daggers and swords. By the time of the American War of Independence it had been in use for over 100 years and established as being used in association with the musket. Other types of weaponry such as artillery were also conventional, but warfare was changed with the development of a crude submersible known as the 'Turtle'. This barrel-shaped vessel constructed from wood and iron was operated by a single man and used to attach explosive charges to British warships as they lay at anchor in harbour. The design was the idea of David Bushnell, an American patriot, and given support to build it by George Washington. On 6 September 1776 it was operated by Sergeant Ezra Lee, who steered it out into New York Harbour to attack HMS Eagle. The action was unsuccessful but during the course of several later operations, the Turtle was used to damage or destroy a small number of British vessels. It was the dawn of a new age, and science and engineering were beginning to change warfare and there would be further innovative developments in weaponry. The Turtle did not directly affect the way in which land battles were fought, but it did show what new forms of weaponry were emerging; but the bayonet would always remain.

Chapter 6

Back to Europe and Other Conflicts

In Europe the bayonet was gaining credibility in the many wars as troops were becoming more and more inculcated in its use. This now meant that on the three main continents of North America, Europe and India, the bayonet had risen from being a useful device for defence and reached prominence where it could be used to force troops away from positions in battle in an offensive manner. On its own the bayonet did not win battles, but in the final stages after artillery had done its job and musket fire further reduced the enemy's rank, the advancing troops with fixed bayonets could clear areas of the battlefield leading to a collapse of resistance. One of the leading Russian military commanders of this age was General Alexander Suvorov, born in 1730 and during his military career fought in many campaigns including the Russo-Swedish War 1741-1743, Seven Years' War, twice against the Turks and other wars. He was an advocate of attacking using 'shock tactics', one of which was the use of the column to mount sudden bayonet charges. He studied the effectiveness of the musket and reached the correct conclusion that the standard service musket was only accurate out to a range of about 60 paces. The average pace is approximately 30 inches, which made his calculation about 150 feet in range.

Suvorov compiled all his theories into a volume and published them under the title of *The Art of Victory*. His ideas formed the basis for a line of thinking which led to an opinion that became known as the 'Cult of the Bayonet'. He believed that his tactic of using small columns to mount bayonet charges violently and quickly over any type of terrain was the way to bring about success. He wrote: 'The bullet misses, the bayonet does not'. A contemporary of his, General Mikhail Dragomirov (1745–1812), who also saw fighting during the Russo-Turkish Wars and later against Napoleon at Borodino, suggested to his troops that: 'If your bayonet breaks, strike with the stocks.' King Karl XII of Sweden, who ruled between 1697 and 1718, was perhaps the most prominent and certainly one of the earliest leading advocates of this opinion in the decisive use of bayonets. At the Battle of Fraustadt in February 1706 during the Great Northern War he managed to defeat a numerically much superior force using aggressive tactics. The Swedish force numbered only 9,000 men to oppose a combined Russian and Saxon force of 18,000 men, but Karl ordered his troops to outflank the enemy and at point of bayonet literally routed them from the battlefield. The Swedish army lost 1,400 killed and wounded whilst they inflicted 15,000 casualties on the enemy.

What Suvorov had seen of the Russian soldier in battle led him to believe he knew what he was talking about when he made a variation on his opinion of the bayonet: 'The bullet is foolish, the bayonet is wise.' The Russian army was largely made up of conscripts from peasant stock such as farm labourers with little if any education. These illiterate troops were taught how to load and fire their muskets as any basic drill movement, but beyond that they had no desire to know or indeed any reason to know more except how to aim and fire. At the end of each day's training, Russian recruits had to repeat the headline words of their military indoctrination by shouting in unison: 'Subordination; Obedience; Discipline; Training; Formations; Military Order; Cleanliness; Neatness; Health; Courage; Bravery; Cheerfulness; Formation Exercise; Victory and Glory' Much of this would have been lost on the peasant stock, who understood little about such things and cared even less, especially when on campaign.

The smoothbore muskets of the period were already inaccurate and in the hands of such poor quality troops, this was compounded by the fact that the sleeve of the socket bayonets attached to the muzzle of the barrel obscured the front sight, which made accurate aiming impossible. Standardization in weapon manufacture was not recognized in Russia and even as late as 1811 when the Russian armouries at Tula alone were producing some 100,000 muskets annually, there was still no standardization in calibre and up to 28 different calibres of weapons were being produced. This made resupplying the troops

Tsar Alexander II inspects Russian troops at Sebastopol.

with the correct ammunition an impossible task. To remedy this deficiency the Russian field commanders maintained Suvorov's maxim and continued to order their troops to advance with fixed bayonets, rather than try to engage with volley fire.

It sounded good but, like the Prussians, the Russians would soon come to realize that such tactics were wasteful in the lives of troops, no matter what their quality. Unfortunately, no army ever learns from the mistakes made by others, it only learns from its own mistakes. The Russians eventually came to learn and changed their tactics, just as Frederick the Great had to discover for himself if the cost in casualties was to be reduced. Major-General Sir R. Wilson, a British army officer sent to observe the Russians fighting the French, reported in 1810 how the: 'bayonet is truly a Russian weapon.' He probably did not realize that the Russians used the bayonet because of poor logistics and the disparity in calibres of muskets. Two years later the manual *Precepts for Infantry Officers on the Day of Battle* instructed how the bayonet charge should be made in deep column formation. Several years previously, the Austrian Archduke Charles believed that troops deployed in the line: 'was the proper formation for infantry, permitting the best use of its weapons, that is the musket for fighting at long range and the bayonet for close in'. The British army knew this and it was the standard tactic, which allowed the whole line to fire at once, whilst the French and Russian column only allowed a narrow front to fire at once. Another manual for bayonet drill recommended that when charging his bayonet: 'a man is firm against any shock, and in guard; shall occasion, or opportunity, to defend himself or annoy his enemy, or to advance against him if he should give way. We have no word of command for pushing the bayonet, the motion being so natural, that one in action can scarce avoid doing it properly.' This motion to 'push' would have been taught in basic training and encouragement to use it on the battlefield would have made it a 'natural' or 'instinctive' movement in battle and requiring no thought.

In eighteenth century Russia under the Tsars, the country was an Empire covering a vast area of land within its borders, but there were other Europeans states such as France and Spain which had overseas territories as part of their respective Empires. England as an island state with a powerful navy to protect its overseas trade routes had also expanded its empire, and this brought it into conflict with European states such as France and Holland. France and England struggled for total domination in India, and in those areas of this vast country which they controlled, they exerted an influence on the way of life of the local populace, including the military and the conduct of war. The local Indian bladesmiths were quick to adopt the European style of bayonet and were soon producing their version of the socket bayonet. Examples of this work still exist in museums, such as the so-called 'Tiger Bayonet', produced in Mysore during the reign of Tipu Sultan (1782–1789). It is very short in length, measuring only 5.25 inches, but it is very heavy and weighs over 1.75lbs, which is much heavier than the designs of European bayonets at the time and these have a much longer blade. The term 'Tiger' was in reference to the familiar term by which Tipu was known – 'the Tiger of Mysore' – after the area in southern India over which he ruled. Local bladesmiths not only produced bayonets for local forces, but they also supplied bayonets to European troops to supplement the standard service issue weapons.

In some armies it was becoming standard practice to carry muskets with the bayonet fitted at all times when on campaign. This meant that troops were in a constant state of readiness to respond to any sudden tactical movement by the enemy, such as an attack by cavalry. Troops in the Prussian Army went one stage further and discarded the scabbards of their bayonets so they had to be fixed all the time. Only when in camp would bayonets be removed from muskets to allow the weapons to be cleaned. Sometimes the bayonets would be fixed to the muskets again in order to allow them to be interlocked, which would allow the weapons to be stacked together in a fashion resembling bundles of stacked wheat. Sentries would patrol the camp area with bayonets fixed because they were on duty and had to be prepared in case of sudden surprise attack. In the event of an attack the sentry would raise the alarm and, as troops emerged from their tents, they could immediately reach for their muskets.

Cavalry could be used to attack any position on the battlefield, but the infantry was still considered to be the most important formation to attack, especially when drawn up in line formation when it was most vulnerable. The defence against such attack was for an infantry unit to assume a formation called a square, which was the universal response against cavalry and historically proven to be the most reliable of all defensive measures against such a threat. The formation itself was nothing new and was a tactic used by the Roman army thousands of years earlier. There were variations on the movement and over the centuries it was also referred to by different terms. The Scottish 'schiltron' or 'sheltron' was an early example of what would become the pike block and was itself evolved from earlier formations. The schiltron was formed by the troops presenting their pikes in such a way that the formation was surrounded by a ring of sharp tips to prevent an enemy from getting close enough to strike. The schiltron was developed through the thirteenth century in battles against the English such as Falkirk in 1298, so that by the fourteenth century it had been perfected to be used effectively in set-piece battles such as Bannockburn 1313, Myton 1319, and Otterburn 1388. It was not always used successfully. For example, the Welsh never quite mastered its use, but when the schiltron formation was taken up by strong troops it was protection against attack by infantry or cavalry. A similar tactic was used by the Spanish army and termed the 'Tercio', which translates to mean 'third' because it was made up of a combination of pikemen, swordsmen and musketeers, and comprised around 3,000 men, which was about one-third the strength of a brigade. The formation provided all-round defence for mutual support and was sufficiently flexible to still allow it a degree of mobility on the battlefield. The creation of the Tercio is credited to General Gonzalo Fernández Córdoba around the turn of the fifteenth and sixteenth century. It was used successfully during the wars against Italy and the idea was even copied by King Sebastian of Portugal. The pike block evolved from these formations and when the bayonet replaced the pike, it was only natural that such a proven formation be retained with some modifications to accommodate the new weapons. The effect was still the same and cavalry units would find themselves faced with an impenetrable obstacle of bayonet points. During the Battle of the Boyne on 1 July 1690 in Ireland, Dutch troops fighting in the army of King William of England formed squares to protect themselves against charges by French and Irish Catholic cavalry, and were able to bring down horses and riders with musket fire and use of the bayonet.

Forming an infantry square is basically a close order formation with each of the four sides arranged to leave a hollow or empty centre into which the wounded could be pulled and the lines adjusted to close up the gaps and maintain the defensive posture. Squares are easier to form as opposed to the circular schiltrons, as troops can be drilled to form this shape from columns or line abreast. Once the danger has passed they just as easily reform their original positions to meet advancing infantry. For well-trained troops who have rehearsed the tactic it is a flexible and strong defensive formation. The smallest number of troops considered possible to form an effective square to withstand a cavalry charge was between 500 and 1,000 men. Each side of the square could be made up of between four and six ranks of infantry with each man armed with a musket with bayonet fixed. The front rank would kneel with the butts of their muskets firmly placed on the ground and held at an angle of around 45 degrees. The centre rank would take aim and fire at the approaching cavalry and the rear ranks would be reloading and getting ready to fire. These ranks would alternate their fire to maintain a steady volley and the front rank with disciplined men presented a formidable obstacle. At the Battle of Waterloo in June 1815 some British infantry squares withstood several cavalry charges or more, and managed to fight them off and hold their positions.

The infantry had to fire at a determined range if they were to stand any chance of breaking up a cavalry charge. Horses cannot be induced to charge into a mass of bayonets; the animals naturally sense the danger as with any other obstacle, but they can be ridden close enough to allow lancers to stab at the infantry and also fire their pistols into the ranks. If the infantry respond and open fire too soon with their largely inaccurate muskets, they will miss their intended target. If they fired too late and brought down the horse and rider close to their ranks, the forward impetus of the horse as it fell might bring it crashing into their positions, where the weight and bulk of the animal could create a gap which could be exploited by other riders. The optimum range to fire at cavalry was around 30 yards (90 feet) if they were to stand any chance of hitting a moving target, and at the same time avoid the wounded horse careering into their ranks.

If a gap was created in the ranks the cavalry would attempt to exploit it and try to 'ride the square into a red ruin' as the formation broke up. The square was at its most vulnerable when it was in the process of being formed and if the cavalry was fortunate enough to catch the troops at this point, the riders would be able to get in amongst the infantry and deal with individuals. At the Battle of Rossbach on 5 November 1757 the Prussian infantry was caught by a force of Franco-Austrian cavalry before they had chance to form squares, but the timely intervention of their cavalry prevented a disaster. A similar incident occurred at Quatre Bras during the Waterloo Campaign in June 1815 when French cavalry caught several groups of British troops unprepared out in the open. However, once the square was formed even the most inexperienced troops could be encouraged to remain at their posts. For example, at the Battle of Lützen on 2 May 1813, French new recruits held their positions to withstand cavalry. So effective was the infantry square that it remained in use as the standard response against cavalry throughout the nineteenth century from Waterloo through the Crimean War and the American Civil War. It is only towards the end of the nineteenth century that the tactic of the infantry square starts to slowly fall into disuse, as advanced weaponry with increased firepower such as long-range artillery with exploding shells begins to appear on the battlefield. By the time of the Franco-Prussian War (1870–1871), with the appearance of modern rifles and multi-barrelled weapons such as the French mitrailleuse, a forerunner to the machine gun, the infantry square was already becoming regarded as obsolete. Despite the advances in weaponry, the manoeuvre was nevertheless still being taught and practised by some countries such as the British army who believed that it was still useful, especially when practised against native troops.

It required discipline if infantry were to hold their positions in squares when attacked by cavalry or bombarded by artillery. For example, at the Battle of Khushab on 17 February 1857 during the Anglo-Persian War, in present-day Iran, Indian cavalry charged an infantry square formed by some 500 Persian troops and smashed their ranks. The Persians did not have the resolve to hold their positions and broke ranks, which allowed the Indian cavalry to get past the bayonets. The result was that only 20 Persians survived the attack. On the South American Continent at the Battle of Acosta Ñu (also known as the Battle of Campo Grande) on 16 August 1869 during the Paraguayan War, a force of Paraguayan troops was attacked by Brazilian cavalry and began to form a square. Unfortunately, the infantry reacted too late and they were caught before they completed their manoeuvre and their ranks were broken up. During this battle between 6,000 Paraguayans and a combined Brazilian and Argentine force of 20,000, there was much use of the bayonet. At the end of the eight-hour long battle, the Paraguayans had lost 2,000 men killed, whilst the Brazilian and Argentine lost only 46 killed and 259 wounded. At the Battle of Isandlwana on 22 January 1879 during the Zulu War in South Africa, the British troops did not form squares even though they were facing an overwhelmingly superior force of some 20,000 Zulu warriors. The British and Native Contingent force of 1,300 men were killed, but it taught the army a sharp lesson which was quickly acted upon. Six months later at the Battle of Ulundi, the 5,200-strong force of British and African troops formed a large open square to face a Zulu force of around 12,000 warriors. At the end of the battle 1,500 Zulus had been killed and wounded at a cost of less than 100 killed and wounded.

Five years later at the Battle of Tamai (also known as the Battle of Tamanieh) on 13 March 1884 during the Mahdist War in Sudan a force of 4,500 British, including troops from the Highland Regiment of the Black Watch, were attacked by a force of 10,000 Mahdists. The British formed themselves into two squares with fixed bayonets. General Graham commanding the square with the Black Watch ordered some of his men to mount a bayonet charge, despite the overwhelming odds. As the men moved forward this left a temporary gap in the ranks, which was spotted by the Mahdists, who surged towards it before the other troops could fill in the space. Attacking en masse, they managed to break into the British square but, being disciplined the men, held fast along with the other square commanded by Colonel Buller. It was a fierce close-quarter action with bayonets being used against spears in a fight during which the British lost 214 killed and wounded, but in turn killed at least 4,000 Mahdists. The following year at the Battle of Abu Klea on the 17 January 1885, 1,400 British troops and Royal Navy ratings were attacked by 13,000 Mahdists. During the course of the fighting a gap opened in one corner of the square and some Mahdists rushed in. Again, the British managed to hold firm and fought off the Mahdists killing at least 1,100 of them at a cost of 75 killed and 82 wounded to themselves. A Gardner machine gun was used during the battle but it had jammed and was out of action for most of the fight. The new machine guns, there had been two Gatling guns at the Battle of Ulundi six years earlier, were ushering a phase of warfare but still the bayonet remained an integral weapon with the infantry and the infantry square despite its waning usefulness was a tactic that would remain in practice to be used in many wars as the nineteenth century drew to a close.

After years of unrest revolution in France in July 1789, the civilian populace, with the support of the army in some cases, rose up to overthrow the unpopular monarchy. The country began to fall into a chaotic state but gradually it became stable and over several years organized into a militaristic state which caused concern in neighbouring states. From 1792 France found itself engaged in a series of wars known as the Revolutionary Wars as it fought campaigns against countries such as Austria, Spain, Germanic States such as Prussia, Italian States and Britain. The army of Revolutionary France used the bayonet often and with strength such as the Battle of Jemappes on 6 November 1792 where the French engaged an inferior force of Austrians of the Holy Roman Empire. One of the French commanding officers Maréchal du Camp Dampierre threw his troops into the fight with vigour and after exchanging volleys of musket fire, his troops managed to take 'the first line of entrenchment with the bayonet'. Other battles followed, such as Haguenau in 1793 and Etigen on 1 July 1796, where the use of the bayonet was also prominent. One of the reasons why the use of the bayonet was encouraged was probably due to the shortage of gunpowder for the muskets caused by the chaos of the Revolution, which disrupted supplies. Sending troops to war who could not fire was no way to win battles, and the responsibility for reorganizing the military fell to Lazare Nicolas Marie Carnot, who would become known as the 'Organiser of Victory'. Carnot was elected to the Committee of Public Safety in August 1793 and appointed military administrator for the French Army. Out of the chaos in the aftermath of the Revolution which deposed King Louis XVI, he endeavoured to create a sense of order and managed to establish a cadre of keen, young, energetic officers to lead the French Army. These were men who had the independence of mind to act on their own accord and use all branches of the army at their disposal to the fullest. In February 1794 Carnot issued general instructions in which the generals were directed that they were: 'always to manoeuvre in mass and offensively; to maintain strict, but not overly meticulous discipline'; he went on to state that they were 'to use the bayonet on every occasion.' This order for 'action with the bayonet on every occasion' was readily taken up by officers such as Captain Elzéar Blaze, who believed: 'To be killed regularly, one had to be killed by the bayonet'. This remark was not meant to imply that one had to be killed frequently, but rather that by being killed by a bayonet it was clean and efficient, which put it in sharp contrast to the remark made during the American War

British infantry circa 1791 with bayonet fitted to musket.

of Independence only fifteen years earlier which claimed that the bayonet was a messy weapon.

During the Revolutionary Wars there emerged an officer by the name of Napoleon Bonaparte who would rise to pre-eminence and had actually been a junior officer in the Royalist Army at the time of the Revolution. He was already recognized as being a man who got things done, but it was during the campaign in Italy during 1796 that his leadership qualities really stood out. In 1798 he led a French army to campaign in Egypt and although initially successful and scored a victory at the Battle of the Pyramids on 21 July that year his army was eventually beaten by the British. Following the defeat France withdrew from Egypt and contented itself to campaigning across the European mainland. During the Egyptian campaign the French had deployed large silk balloons filled with hydrogen gas to allow observers to see over the battlefield and drop messages to the ground to give details of what the enemy were doing. These were part of the unit called the 'aerostatiers' which had also been present two years previously at the Battle of Fleurus in June 1794. Despite great hopes for these devices, they proved inadequate and Napoleon was not convinced of the practicalities of the observation balloon. Furthermore, they could take between 36 and 48 hours to inflate, which rendered them limited in operational terms and a battle could be over by the time they were deployed. Science was conquering another dimension by enabling flight, but to the infantryman on the ground with his bayonet this meant very little.

In 1805 Colonel (later Sir) William Congreve of the Royal Artillery showed another scientific advance when he demonstrated his newly developed exploding rockets to a gathering of military observers at the Royal Arsenals in England. Rockets dated back many centuries, the Chinese had used them in warfare in the thirteenth century, and Congreve's designs were, in fact, based on a design used by Indian troops of Tipu Sultan during the Mysore Wars. Colonel Arthur Wellesley (later the Duke of Wellington) witnessed the rockets in action but was never a keen believer in the weapon. Scientific research improved the design and the British army used rockets on several campaigns and the Royal Artillery established special batteries of rockets. Wellesley was a conventional soldier who believed in the value of his troops and the traditional weapons they used, which included the bayonet. He had developed this belief through many engagements such as the Battle of Assaye on 23 September 1803, during the second Maratha War, when he commanded a force of 9,500 troops of the East India Company to face an army of 40,000 Indian troops, including 11,000 of which were trained in the European methods of fighting. Despite being vastly outnumbered Wellesley's troops fought well, especially the 78th Highlanders and sepoys who used their bayonets to good effect and got 'at the enemy' with volley fire followed up by pressing in with the bayonet. Indeed, it has been said that India was a continent won for the British Empire by point of bayonet. Wellesley's force lost 1,600 men killed and wounded and inflicted 6,000 killed and wounded on the Indian force. He would later say that Assaye was his greatest battle and one of his hardest fought.

Between 1802 and 1814 Napoleon's skill won him many battles and Wellesley would command and lead the Allied forces to victory in the Peninsular War in Spain and Portugal against the French. This was a war in which the bayonet became valued and its use in battle was often expressed such as at the Battle of Corunna on 16 January 1809 when the Scottish regiment of the Black Watch misunderstood orders to replenish their ammunition and instead began to withdraw. The British commander Lieutenant–General Moore seeing this movement rode up and addressed the men personally by declaring: 'My

British infantry charge with bayonets at Talavera.

brave 42nd, if you have fired your ammunition, you still have your bayonets!'
The British use of bayonets to force a battle was exhibited five months later
at the Battle of Talavera on 27–28 June south-west Madrid. At this action a
combined Spanish-British force of over 55,600 men routed a French force
of more than 46,000. The British troops in particular resorted to bayonet
charges to break up the French lines. In the end the British lost 6,268 killed
and wounded and the Spanish 1,200 killed and wounded. The French lost
7,389 killed and wounded, each one of which had to be replaced by a new
recruit along with all his equipment.

It was inevitable that two such eminent military commanders as Wellesley
and Napoleon were bound to come into direct conflict sooner or later as
they followed parallel careers, but that event would not happen until 1815.
Wellesley knew of Napoleon's reputation and said that his presence on the
battlefield was worth 40,000 men and how even in defeat he was able to survive.
After a number of campaigns and many victories across Western Europe
Napoleon Bonaparte suddenly changed tactics from his pan-European war
and in June 1812 he attacked Russia. It was to be his greatest challenge and
Russia would turn out to be his most determined adversary. The Russians
had a predilection to the use of the bayonet, a fact the French seemed to
have forgotten since the Battle of Morungen in 1806 in modern-day Poland
when Russian troops of the Yekaterinoslav Grenadier Regiment showed their
contempt of French troops by declaring they were not worthy of having
bayonets used on them, and instead charged the French positions using the
butts of their muskets. Napoleon finally reached Moscow in September but
the city was virtually deserted and there were no stocks of food to feed his
troops. When the city caught fire he was forced to abandon his occupation
of the city and withdraw. Winter was settling in and the long trek back to
France was a journey laced with danger at every footstep. They were attacked
all the way and the retreat became a series of rearguard actions. For example
at the Battle of Maloyaroslavetz on 24 October 1812 the Russians kept
up the pressure by harrying the French and a Russian NCO was heard to
shout: 'shoot at everything French, and keep up the scare with bayonets'.
The Russians maintained the pressure and an officer by the name of L.A.
Shimanski later recalled in his memoirs how he: 'saw, for the first time how
they [the French] were punished by bayonets'. This was how bayonets were
to be used in battle by making a retreating enemy hasten their departure from
the battlefield. At the later Battle of Kulm in 1813 a French force was caused

to fall back in retreat in the face of stalwart Russians advancing with fixed bayonets. The French were caught up in a crush as they tried to cross a stream in an effort to escape the Russians. They became a disorganized mass and the Russians fell upon them with a terrible fury.

Following the disastrous campaign against Russia in 1812 when he lost an entire army, Napoleon could still raise new recruits to continue France's wars. He was known to hold a keenness for the use of the bayonet and even devised a series of strategies and tactics, which were developed to allow his troops to use their bayonets en masse with maximum effect. One of his tactics was known as the 'ordre mixte', which had been devised by Count Guibert with battalions deployed to support one another and used weight and strength to engage the enemy, but firepower and the threat of the bayonet also added impetus to the tactic. According to Captain Blaze, Napoleon 'had a certain fondness for those who perished in this manner'; not that Napoleon would have personally known too many of his troops who were bayonetted. At the Battle of Eylau fought over two days between 7 and 8 February 1807, a total of eight senior officers were killed and fifteen wounded including Chef de battalion Jouradet, who was wounded by a bayonet thrust which would have made him worthy of attention by his commander. Napoleon also had a healthy respect for the

use of the bayonet, which stemmed from personal experience having been wounded by a bayonet thrust from a British sergeant whilst he was leading an attack during the siege of Toulon on 17 December 1793. It was undoubtedly painful, but he was fortunate the wound was to his thigh and could be treated easily with minimum risk of infection and little loss of blood, unlike an abdominal wound. It would be the only wound he ever sustained in battle.

Throughout his military career Napoleon relied on the proven use of massive attacks using large columns and even divisional formation. At the Battle of Austerlitz on 2 December

Napoleon Bonaparte was wounded by a bayonet and had a high regard for the weapon.

1805 the use of the bayonet by French troops delivered the Pratzen Heights, a feature of the battlefield, to Napoleon and gave him a victory. During this battle the Russian Guards made a courageous bayonet charge covering 300 yards and broke through the first ranks of the French positions. However, they were exhausted by their efforts and the French were able to counterattack and drive them back. Things were much different seven months later at the Battle of Maida on 2 July 1806 French infantry of the 1st Léger Regiment faced British troops and both sides advanced without firing until separated by 100 yards of open ground. Both sides then opened fire and advanced at point of bayonet. On this occasion it was the British troops who advanced with more determination and the French recoiled and withdrew. The push of the pike from the seventeenth century had now been replaced by the push of the bayonet and for the same purpose. The bayonet could be used by a large number of troops to remove an enemy from a position by using the 'push of the bayonet' or by a single man for self-defence or to provide a rearguard action. For example, at the Battle of Busaco in Portugal on 27 September 1810, a 19-year-old soldier by the name of Stewart who was serving with the 43rd Regiment (later to become the Oxfordshire Light Infantry) faced down French adversaries. Despite his youth the young man was apparently of large muscular build, which earned him the nickname of 'The Boy'. As his unit crossed a bridge being pursued by the French he turned to face them and engaged in bayonet fighting. Charles Napier, later to become General Sir Charles Napier, was a young officer who observed the incident and remarked how: 'Then, striding forward in his giant might, he fell furiously on his nearest enemies with the bayonet, refused the quarter they seemed desirous of granting, and died fighting in the midst of them.' It was a selfless act of defiance which helped forestall disaster. It was also an act which added to the legend of the bayonet the history of which was becoming peppered with many such acts.

Another French officer of the time who believed in the bayonet was the historian Colonel Gay Vernon of the École Polytechnique, the author of *Treatise on the Science of War and Fortifications* published in 1817. He believed that it took only the first volley of musket fire to decide the outcome: 'after which the bayonet and the sword may charge without sustaining great losses' That would certainly appear to have been the case at the Battle of Maida but there were other battles where it took more than the threat of the bayonet to win the battle. Between 1792 and 1815 it has been calculated that the French armies of Napoleon fought 611 major engagements and between 1802 and 1814 Napoleon's forces certainly fought one army after another from Russia in the east to the Spanish and Portuguese coast in the west. His troops and those of his enemies used their bayonets in such famous battles as Borodino, Jena, Eylau and Dresden. The French army may have made use of the bayonet but other armies also recognized its value on the battlefield when it was used en masse at an optimum moment. The Austrian and Prussian armies stressed the use of the bayonet during an attack and the Russians particularly appeared to relish its use. There was still the odd occasion when men had recourse to resort to the use of the musket butt as in times of old. For example, the weather conditions at the Battle of Katzbach on 26 June 1813 were so terrible and the rain so heavy that the powder in the touch-holes of the muskets was soaked and would not ignite under the spark of the flint. Prepared cartridges were also soaked and reloading muskets was impossible. The Prussians were not stalled by this and attacked the French using the age-old combination of bayonet point and butt. The combined Russo-Prussian force of 80,000 lost 4,000 killed and wounded, but they inflicted 12,000 killed and wounded on the French force of 60,000 and gained a victory. On the same day Napoleon was engaged in fighting a battle at Dresden where his army of 100,000 men defeated an Austro-Russian force of 200,000 men. Armies were becoming larger and weapons were becoming more powerful, and this meant battlefields also increased in size to accommodate the thousands of additional troops. The nature of war also meant that weapon production had to be increased to an industrial scale from pistols to muskets and of course bayonets along with other bladed weapons such as swords.

Napoleon scored what would turn out to be his penultimate victory at the Battle of Brienne on 29 January 1814 against a relatively small Russo-Prussian army, but he

Napoleon's return from Elba, 1815.

was beaten three days later at the Battle of La Rothière by a much larger Austro-Prussian force. The British victory at Toulouse on 10 April that year was the final disaster and after many campaigns Napoleon was left with no option but to abdicate and in May that year he was sent into exile on the island of Elba allowing peace to return to Europe. It was to be a short-lived peace, though, because on 26 March 1815 he escaped from the island with a small group of soldiers. He landed in France, near Cannes and reached Paris on 20 March having recruited an army along the way. The Allied coalition force headed by Britain declared war on him and Napoleon began his campaign by invading Belgium. The two sides first came into contact on 16 June 1815 at Quatre Bras and Ligny on the same day where Napoleon beat the Prussians in his last victory. At one point during the Battle of Ligny the Prussians did succeed in driving the French from their positions at the point of the bayonet, but Napoleon won the day and inflicted 20,000 casualties on the Prussian army for a loss of between 6,000 and 7,000 killed and wounded. Two days later the Allied army under command of the Duke of Wellington faced Napoleon on the battlefield for the first and only time at Waterloo. The first shots of the battle were fired at around 1.30pm and the fighting began in earnest at around 2pm on 18 June. The battle can be categorized in six main phases with fighting lasting until darkness began to fall at 7.30pm.

Many incidents occurred over the three days of the Waterloo campaign, some involving individual acts of use of the bayonet. For example, at the Battle of Quatre Bras, Ensign Christie, carrying the regimental colours of the 44th Regiment of Foot, was attacked by a French lancer who attempted to wrest it from the unfortunate officer but he was bayonetted for his efforts. Infantry on both sides formed squares to present rows of bayonets to counter cavalry charges such as the 6,000 British troops of Alten's Division who arranged themselves

French infantry fire Charleville muskets.

into nine separate squares. French cavalry could not penetrate the ranks and although some riders approached they were kept at bay. Statements after the battle recall how mounted cavalry could not charge because of the bayonets and the infantry could not reload because that would allow the cavalry a chance to attack. The result was a standoff and each side looking at the other and not able to do anything about it. A Royal Engineer taking shelter in the square formed by the 79th Regiment of Foot recalled how French cavalry acted: 'No actual dash was made upon us. Now and then an individual more daring than the rest would ride up to the bayonets, wave his sword about and bully; but the mass held aloof, pulling up within five or six yards, as if, though afraid to go on, they were ashamed to retire. Our men soon discovered they had the best of it, and ever afterwards, when they heard the sound of cavalry approaching, appeared to consider the circumstance a pleasant change (from being cannonaded)!' This impasse between cavalry and infantry could not last very long in the middle of a battle and the infantry square certainly was not going to move and risk breaking ranks and thus lower its protection from the defensive rows of bayonets. It was always the cavalry that rode off to search for more vulnerable targets.

Recreated bayonet defence to show part of a 'square' formed against cavalry.

One of the most outstanding single actions on the battlefield of Waterloo that day was that of Sergeant Charles Ewart, later Ensign, who was serving with the Scots Greys as part of the Union Brigade and as such was engaged in action at close quarters. Ewart was an expert horseman, extremely strong and stood well over six feet tall. He was also a proficient swordsman as befitted a cavalryman. He managed to capture a French regimental 'Eagle' which became a fight for his life. He later recalled his actions in a letter home to his brother in Ayr on 2 October 1815: 'It was in the first charge I took the eagle from the enemy; he and I had a very hard contest of it. He thrust for my groin; I turned it off and cut him through the head; after which I was attacked by one of their Lancers, who threw his lance at me, but missed the mark by my throwing it with my sword by my right side; then I cut him from the chin upwards, which went through his teeth. Next I was attacked by a foot soldier, who after firing at me charged me with his bayonet, but he very soon

1st or Grenadier Regiment of Guards, circa 1815.

lost the combat, for I parried it and cut him down through the head; so that finished the combat for the eagle.' The cavalryman had the advantage over his opponent in height but the infantryman would have been more agile. Ewart was fortunate to come away from the encounter because, had there been more French infantry, they would surely have overwhelmed him.

The British positioned troops at the farm of Hougoumont on the right of their positions and at La Haye Sainte in the centre. The troops used their bayonets to cut loopholes in the walls through which they could fire their muskets. As the French attacked these positions they made desperate attempts to dislodge British and grabbed at the bayonets as they protruded through the walls. The battle continued to rage all round and although the French did manage to enter these positions the British held firm and engaged in close-quarter fighting using bayonets. Finally, the French had exhausted every means at their disposal and in a last act to try and wrench victory, Napoleon ordered an attack to be made by the elite Imperial Guard at around 7pm. They marched forward bravely shouting 'Vive L'Empereur' and 'En avant a la baionnette' but it was not sufficient to overcome the massed firepower which awaited them as they strode up the hill. The British troops which had, until then, been lying down suddenly stood up, presented arms and fired. The French recoiled as volley after volley crashed into their ranks. General, later Sir, Peregrine Maitland commanding the Foot Guards ordered a bayonet charge. The French troops broke and began to disperse being pursued by the British. A second column of the Imperial Guard began to advance and Maitland's troops were forced to retire to their original positions where they reformed. The second French column was assailed by intense musket and artillery fire, which forced them to retreat. The battle was finally over and in the last moments the bayonet had proven effective in breaking up enemy lines.

Chapter 7

New Weapons and More Wars

After all the years of fighting across Europe, the peace which returned to the Continent after the defeat of Napoleon, came as a relief, but especially to Britain which could now release the Royal Navy to other duties such as protecting trades routes, and the army was free to help police the country's growing Empire. European countries still bickered among themselves with various territorial disputes. Some of these arguments erupted into full war while others were no more than border clashes. This meant that while European countries were still fighting in the traditional styles with large armies drawn up in linear ranks with all branches of the army represented on the battlefield, the British army was engaged in colonial wars against local forces usually of poor quality but inspired by religious fervour in some cases such as the Mahdist War (1881–1899), in Sudan and Egypt. They were often poorly equipped, but the armies such causes could muster hoped to win by force of numbers.

Britain would find some of these tribal forces were easily subdued but others would be a force to be reckoned with, such as the Zulus in 1879, the Ashantis and the Mahdists. It was also a period of great development which turned into an arms race with very powerful weaponry being produced. This was a period referred to by the German historian and philosopher Oswald Arnold Gottfired Spengler as: 'war without war, a war of overbidding in equipment and preparedness, a war of figures and tempo and techniques.' Artillery and muskets were now being developed into breech-loading designs, which increased their firepower and the lethality of the projectiles they fired with greater accuracy. Developments in other areas included submarines and more powerful warships mounting large-calibre pieces of artillery. This was an international arms race on an industrial scale. The first practical breech-loading weapons were beginning to appear but the bayonet as a weapon still remained in service with these new rifles although the design would change to suit the requirements of each individual army.

In America troops were involved in dealing with local natives who had no technology and fought using any firearms they could obtain, otherwise they fought using bows and clubs which had not changed in thousands of years. The army having superior weapons were usually able to keep these natives at a distance and there was little need for bayonet fighting. Indeed, the bayonet appears to have virtually disappeared from military use in the American army and would return in the latter half of the nineteenth century and come to great prominence during the Civil War. American troops fought in the Mexican War with engagements such as the defence of the Alamo in 1836, which pitted them against a properly equipped modern force in Texas but never an army outside the country. In Europe, by contrast, countries such as Austria were engaged in supressing a revolt in the Papal States in 1831 and Prussia and Denmark fought a war in 1848–1849 over the sovereignty of the region of Schleswig-Holstein. In the meantime, the British army fought a hard war against native Maori warriors in New Zealand and in India many battles there proved how formidable native troops could be. Artillery had improved, but it was the infantryman's personal weapon which was transformed from being an inaccurate smoothbore weapon to a type fitted with a rifled barrel for greater accuracy; and when fitted with a magazine for brass cartridges it was turned into a multi-shot weapon. Naturally, these changes also meant that the bayonet had to change in order to fit the new weapons. Until these changes were completed, the socket bayonet with its triangular-sectioned blade with its long tapering form remained in widespread use.

Changes in firearms had begun as early as 1822 when London-based gunmaker Joseph Egg and his contemporary Joseph Manton started an evolution in weapon design that would lead to the revolver, which perfected by American gunmaker Samuel Colt. These developments led to the production of the multi-shot pistol. Developments in ammunition led to design on the brass cartridge case with a fixed bullet head, which replaced the paper cartridges and, with an integral primer in the base, these metal cartridges were fired by a long metal pin. The British army adopted the Brunswick rifle into service in 1835 and other armies followed suit. The Prussian army adopted the rifle developed by Niklaus von Dreyse known as the 'zundnadelgewehr' or needle ignition rifle. The first version entered service in 1841 as the 'Leichtes Perkussiongewehr Model 1841' light percussion rifle Model 1841. It served the Prussian army well through various conflicts including the Austro-Prussian War of 1866 and the Franco-Prussian War 1870–71, after which it was replaced in service. It was fitted with bayonets which were fitted with spring-loaded clips in the handle, and the blade developed into a flattened sword-like design. The French army accepted into service the Minié rifle followed by the Chassepot, which served throughout the Franco-Prussian War, followed by the Gras rifle and the Lebel. The British army accepted first the Snider followed in 1871 by the Martini-Henry and naturally all these new designs had to have new designs of bayonets. The role of the bayonet remained essentially the same, which was to provide defence and also allow the infantry to go on the offensive, but the main threat now came from massed attacks by native forces rather than cavalry.

France adopted bayonets with exceptionally long blades known as sword-bayonets, a trend followed by a number of other armies, including the British army. One particular design was attached to the rifle by passing the muzzle of the barrel along the length of the handle to fix into the guard. This particular design measured an incredible 3 feet and 9 inches and yet only weighed 1lb 11oz. The weapon was developed in the mid-nineteenth century and was essentially a cavalry-style sword, which made it ideal to wield as a hand-held weapon. When attached to a rifle it would have provided an advantage in having a greater reach when lunging, but it would have been ungainly in length,making it clumsy and, although intimidating, it would have been unsuited for close-quarter fighting. Other French designs for sword bayonets were of a more conventional and useful length such as the 'Baionette sabre' (sabre bayonet), used on the M1837 carbine and M1838 rampart gun. The blade for this was offset rather than being in straight line. The socket-type sword bayonet for use with the Perrin rifle was more traditional and the blade was formed into a style referred to as 'Yataghan', which had a slight curve to the length of the blade. This type of blade was retained for the bayonet used on the M1866 Chassepot rifle and provided a long reach for lunging at a target, and the style also influenced several other armies.

The two main types of blades used on bayonets were originally either flat dagger-like or triangular in section, but over the centuries there have been some curiosities in styles, examples of which are held in museums today and even sought after by collectors. The triple-blade style was never introduced into service but some blades were actually adaptations of local knife styles. One such location where this happened was the Indian sub-continent, which became the battleground between France and Britain as each sought to gain control of the enormous wealth the country obviously offered. The vast country had always been a centre of weapons production and by the eighteenth century bladesmiths there were producing a range of locally made bayonets. These were hand-forged socket-type bayonets but the blade was greatly reduced in length, being approximately half the length of the European designs. The blades of these weapons were usually a flat design as opposed to the European trend for triangular-section blades, because this was the easiest shape to produce, but that did not affect the ways they were used in close-quarter fighting. Local native troops raised to serve in the British or French ranks would have been issued with either locally produced bayonets or European style bayonets, to fit the service firearms such as the French Charleville musket or British Brown Bess.

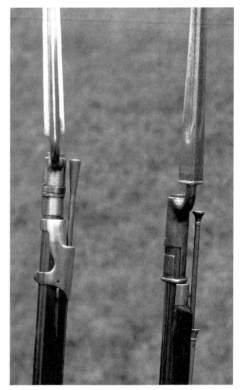

Charleville left Brown Bess right. Notice the difference in the shape of the ramrods.

One particularly belligerent local force to emerge was the warrior-caste of Nepalese from the Gurkha Kingdom, who first came into contact with British troops when the East India Company fought the Gurkha War between 1814 and 1816. So impressed by the Gurkhas' fighting prowess was the East India Company that in 1817 they were asked to serve in the company's ranks as troops. The Gurkhas, according to some sources, were so impressed by the steady use of the bayonet in battle by the British and Indian troops that they agreed to the offer. Gurkha troops later took part in the Anglo-Sikh Wars of 1846 and 1848, and during the Indian Mutiny of 1857 they were most loyal of all native troops serving with the British. This loyalty continues to this day and follows on from the staunch service they gave through both world wars. In 1982 they were part of the task forces sent to recover the Falkland Islands after the invasion by Argentina and served alongside Welsh and Scots Guards, Parachute Regiment and Royal Marines. Their traditional fighting knife, which also serves as a tool, is the heavy-bladed Khurkuri – a term which has been 'westernized' over the years to be spelt as 'kukri' and pronounced as such. It has a distinctive forward curved blade and is heavy enough to cut through sinew and muscle tissue and is still carried today. In an effort to equip the Gurkha troops with a weapon which was familiar to them, they were issued a bayonet which incorporated the Kukri-style blade. It was an unusual item which saw the standard socket bayonet sleeve being fitted with a curved kukri blade. The exact origins are still not precise and some sources claim it was devised by the East India Company. This would seem the most logical point of origin given the direct contact. The bayonet probably appeared around 1820 and although there were slight variations in the size the average overall length was around 17 inches and a blade length around 12 inches. Looking at the arrangement it would have been a clumsy effort and some bayonet collectors believe the kukri bayonet was only ever intended for ceremonial duties and that the standard socket bayonet would have been issued for battle. Indeed according to a spokesperson from the Gurkha Museum at Winchester in Hampshire, England, the kukri–style bayonet 'did not have any practical fighting function.' It was apparently put to better use as an item for display on ceremonial occasions and parades.

The traditional style blades for bayonets were most commonly used armies because the design meant they could be mass-produced easily, cheaply and quickly because of their shape. There were those who believed that the wider the blade the greater the wound and they were, of course, correct, but there is an optimum even in blade size for bayonets. However, that did not prevent some very unusual blade types from being designed, which were more expensive to produce because of the time involved and the limited numbers required. For example, the Spanish produced a style referred to as a 'halberd' bayonet from its resemblance to the medieval polearm of the same name, and fitted with a crescent-shaped axe-like cutting blade to the quillon on one side. The other quillon was elongated to produce a spike and the blade was a flamberge design with its undulating wavy blade extending forward. The overall length of the bayonet measured 17 inches approximately and weighed 1.75lbs. It was fitted with a brass hilt and was the idea of a Spanish businessman by the name of D. Juan Aldasoro Uribe, who proposed it would be fitted to the M1857 carbine. This weapon was already capable of being fitted with a standard bayonet, but Uribe believed his design could be used by the Royal Halberdier Guards. It was a short-lived idea as in 1868

General Bhimsen Thapa surrenders to the British at Sugauli, 1814. The Gurkhas are armed with kukri bayonets. (Collection of Lok Bhakta SJB Rana. Supplied by the Gurkha Museum in Winchester)

the Halberdier Guards were disbanded. Examples of the halberd bayonet are now on display in museums and collectors believe the style may have been for ceremonial purposes, because in the nineteenth century this style was totally impractical for the type of fighting at the time.

In February 1871 the British Small Arms Committee passed for service with the army a new style of bayonet which measured 20.75 inches in length and weighed just over 1lb 6oz with the front end which swelled out to a spear-like shape. This was the 'Elcho' pattern and it had a spring-loaded locking bolt in the handle and a muzzle ring on the quillon for secure fixing and the back of the fullered blade was serrated for sawing wood, a feature which was also beginning to be used by other armies. It was issued to the 23rd and 42nd Regiment of Foot (later to become The Royal Welsh Fusiliers and The Black Watch (Royal Highlanders) respectively) along with the 2nd Battalion the Rifle Brigade for use on the Snider rifle. The combination was used during the Ashanti War of 1873-1874 but the design did not come up to expectations and was discarded. The weight of the bayonet made it useful as a hand tool for chopping wood and clearing away scrub but as a thrusting weapon it was 'indifferent'. Lord Elcho was born Francis Charteris in August 1818 and became the 10th Earl of Wemyss on the death of his father in 1883. He commanded the London Scottish regiment for 17 years after it was formed in 1859 and enjoyed an interest in many things, one of which was designing the bayonet which bore his name. Some sources state the Elcho bayonet weighed 1lb 7oz and measured 25 inches and this may have been a variation, but the fact remains it never attracted much in the way of support from the army, and the design joined the list of those types of bayonet which failed to generate any interest.

In 1885 Lord Elcho proposed his bayonet design for service and stated the: 'desirability of substituting the sword-bayonet invented by me for the present triangular-bladed weapon.' He was nothing if not persistent in his attempt to get his bayonet design accepted. At the time the British army was conducting a series of campaigns in Africa where the undergrowth had to be cut away using tools such as billhooks and Lord Elcho (Wemyss) had the idea to use his bayonet to provide such a cutting tool and bayonet all in one. Not all agreed with him as to the merits of his design as an item which appeared in the United Services Gazette outlines by reporting that: 'Lord Wemyss conjectures that the proportion of billhooks to men is very small and that his sword-bayonet would have been most useful to our men at Suakim, who were largely employed in clearing the bush. We are sorry we cannot agree with him. Recent experience has shown that, notwithstanding the statement of Lord Wolseley with reference to bush-clearing, that the days of hand-to-hand fighting are by no means a thing of the past, and that the soldier has found in the last 13 months a "prodding" weapon [to be] of great value... No, the bayonet and the billhook cannot be combined, and any attempt to do so will only result in giving the army a bad thrusting weapon and an indifferent cutting tool.' The Elcho bayonet continued to be part of experiments with other firearms such as the Martini-Henry rifle along with first the Lee-Metford and then the Lee-Enfield bolt-action rifles. It underwent a number of design changes but finally it was discarded for good.

The major political event of the time happened in France in 1848 when the nephew of Napoleon Bonaparte, Louis-Napoleon Bonaparte, was elected President of France and three years later in a coup d'état he was established as Emperor Napoleon III, a position he would hold until 1870 when the Franco-Prussian War was going badly for the country. Under his sovereignty France gained new colonies and engaged in a number of overseas campaigns including Mexico, the Far East and North Africa. Napoleon III sent troops to support Turkey during the Crimean War and in 1859 he sent troops to help in the unification of Italy. This move brought France into conflict with Austria, which resulted in battles such as Magento and Solferino on 24 June 1859. At this battle some 138,000 French and Sardinian troops were deployed to face an Austrian army of 129,000. Among the French ranks were colonial troops such as Turcos from North Africa. These troops showed a willingness to close with the bayonet and during the battle rushed headlong into Austrian positions. A newspaper account of the action reported how positions were overrun and mentioned: 'These small enclosures had to be carried at the point of the bayonet. I saw several of them which were literally covered in dead bodies. I have counted more than 200 in a small field, not 400 yards in length by 300 in breadth' Several more bayonet charges were conducted before the battle ended with a French victory. The French had lost over 1,600 killed and 8,500 wounded. The Sardinians lost over 4,000 killed and wounded. The Austrians lost almost 2,400 killed, over 10,600 wounded and more than 9,200 missing. Swiss-born Jean-Henri Dunant heard of the battle and visited the scene and on learning of the terrible carnage he set about establishing the Red Cross to care for the wounded in conflict.

Other countries across Europe were also coming into conflict seeking to break away from outside influences or domination. Between November 1830 and October 1831 Polish troops rose up in the so-called 'November Uprising' to rebel against Russian dominance. In the almost year-long war some 150,000 Polish troops were involved in the fighting against up to 200,000 Russians. Around 40,000 Poles were killed and the Russians lost 23,000 killed and perhaps 60,000 wounded. One of the last actions was fought around the Wola garrison on the Vistula River near Warsaw. On 6 September 1831 during the battle the Russians exhibited their vigour for using the bayonet when they engaged a force of some 3,000 Poles. They charged with fixed bayonets and entered the Wola redoubt, a fortified position, and entered the site. One of those known to be bayonetted to death was the commanding officer General Josef Sowinski who was a veteran of the Napoleonic Wars and had fought at the Battle of Eylau. Records state that after the fighting had ended there were only eleven Polish troops alive. This incident and Napoleon's wounding by a bayonet proves that officers were exposed to being killed or wounded by bayonets equally as any other rank. In that sense bayonets follow the path of all other weapons on the battlefield and are no respecter of rank, status or role. In close-quarter fighting if the enemy gets in the way he will be either shot of bayoneted.

As the nineteenth century progressed so a new class of military theorists emerged whose writings revolutionized the ways in which wars were fought such as the Prussian officers Clauswitz, 1780–1831, and Friedrich von Bernhardi, 1849–1930, and French-born Ardant du Picq, 1821–1870 and his fellow countryman Baron Antoine-Henri Jomini, 1779–1869, who would later serve in the Russian army. Each of these men in turn knew that warfare was changing and battlefield tactics had to change to accommodate new weapons with greater firepower such as the Prussian army's breech-loading Dreyse rifle. During the Austro-Prussian war of 1866, also known as the Seven Weeks War from the length of its duration from 14 June to 23 August 1866, Austrian officers still ordered units of infantry to advance across the battlefield and expected them to capture positions with point of bayonet. It was a strategy which would cost them dearly, as in the Battle of Koniggratz (also known as Sadowa) on 3 July 1866 where they lost 41,000 men killed, wounded, missing or captured largely due to old-fashioned tactics. Ardant du Picq attained the rank of Colonel in the French army and saw service in the Crimean War, France's colonies in North Africa and was mortally wounded at the Battle of Borny-Colombey on 18 August 1870 during the Franco-Prussian War of 1870. Like his Prussian counterparts he left behind a legacy of written military theories including his work *Etudes sur* la Combat, which was not fully published until after his death. He understood that warfare was changing had witnessed weapons develop from single-shot muzzle-loading muskets to multiple-shot magazine rifles.

Ardant du Picq sought to impart the tactic that advancing infantry should continue to fire as they moved forward and believed that 'he will win who has the resolution to move forward.' To some this still meant using the bayonet, but there was more to his ideas than that. Ardant du Picq had realized that firepower was all-important but it was the final moments when closing with the enemy using bayonets that could make all the difference, as had been shown many times; but the cost in casualties was always high. He wrote that: 'Each nation in Europe says: 'No one stands his ground before a bayonet charge made by us'. And all are right.' Many armies believed the bayonet to be 'theirs' by heredity and held firm that by its use they would win through. Some may use it as an act of defiance but few actually stood firm in the face of a determined force advancing with fixed bayonets. As tactics changed and

new weapons were developed
the thinking of officers had to
change, but there were some who
held onto the traditional ideals.
One of these 'old traditionalists'
was General Sir Henry Gough
who on being informed at the
Battle of Sobraon in India on 10
February 1846 that the artillery
was running short of ammunition
replied: 'Thank, God! Then I'll
be at them with the bayonet.' He
may have been able to say such
things against an Indian army
but against a European army
with modern weapons such an
opinion would have cost him
dearly. His line of thinking may
have been influenced by the
actions from only two months

Victorious Prussian troops march through the
Arc de Triomphe in Paris in 1871.

earlier where troops of the East India Company had faced down the Sikh
guns at the battle of Mudki and Firozshah fought on 18 of December and
21-22 December respective. At these actions it was reported how; 'With the
indomitable courage that has ever made their countrymen proud of them, the
British troops advanced. Resolute and reckless of everything but the victory
they were determined to win, they carried battery after battery at the point
of the bayonet, surging on with the unimpeded majesty of a storm cloud,
and rolling rank after rank over and through the Sikh encampment' It was
thrilling stuff for the British public to read in the popular press and similar
reports appeared in journals in other countries. To the soldiers in the field it
was a much different matter as they faced death in battle or through disease
and although they won against Indian troops the outcome would have been
much different against a European army armed with modern weapons.

The second half of the nineteenth century would prove to be a period of
great change brought about by conflicts such as the Crimean War, Indian
Mutiny, American Civil War and numerous other wars fought in locations
around the world. During this time infantry weapons changed greatly and

firepower increased. But they all had the one common denominator in the form of bladed weapons such as swords but especially bayonets. The range which weapons could fire out to had increased dramatically but infantry still had to advance to contact to decide the outcome and force the opponents off the field. It was also a time when the style of the bayonet was moving away from the traditionally accepted long tapering blade with its triangular cross-section shape. These were still socket-type fixings with locking rings but some types such as the sword-bayonet for the Baker rifle showed how the bayonet could be mounted offset and fitted to a bar or 'lug' which fitted into the handle.

The socket-type fixing was still very popular because it offered the simplest form of attaching a bayonet and the British army still used the design for the bayonet issued for use with the Sappers and Miners Carbine in 1841. This was an experimental type with a flat blade measuring 25 inches in length and weighing 2lb 3oz . The blade had a saw back edge to help in cutting wood, but this design would later lead to allegations of which had a more gruesome use as all armies accused one another of war crimes and other atrocities because of its pattern. This layout was later modified but the bayonet never appears to have entered service on a wide scale. A similar bayonet was introduced for the Sappers in the British East India Company, but it lacked the saw back. In 1837 the French introduced a flat blade 'Baionette Sabre' fitted with a detachable brass handle, which allowed it to be used for chopping wood, but removing the handle allowed it to be fitted to a rifle as a traditional socket-type bayonet. In 1865 the French introduced another flat blade design sometimes known as the Yataghan, which was in effect a double curved style, and which would remain in service for use on the Chassepot rifle. Twelve years later the French introduced the M1874 'Epée' bayonet, which had a long, slim tapering blade and used with the Gras rifle. This was the type which would become called the 'Rosalie' and serve through the First World War and continue until the Second World War. The spring-loaded catch in the handle locked onto a stud on the barrel and one of the quillons was flattened out with a circular cut-out, which fitted over the muzzle of the barrel to provide a more secure fixing and prevent lateral movement. The other quillon was bent forward like a hook, and was intended to catch an opponent's bayonet blade in close-quarter fighting. It was a style which would be copied by other armies such as the British and Japanese. The Russian army kept the socket-type of fixing which was still being used on the M1891 Mosin-Nagant rifle, but the blade was a distinct, indeed unique, cruciform shape in cross section. This type of blade pattern would continue to be used until the Second World War and would be used to provide forensic evidence of an atrocity committed against Polish soldiers at Katyn.

The flat blade style and spring catch in the handle was now becoming universally accepted and over the next century or so it would be the length of the blade which would vary to alter the shape. Some types of bayonet could have blades the same length as the original socket-types but others would reduce in length to a more manageable knife or dagger-type size with a blade measuring no more than several inches. Indeed by the time of the Second World War in 1939 German and American bayonets were of this type and retained the dagger-like shape which would continue into post-war years and now used on modern rifles. The Japanese kept to bayonets with very long blades and at first the British army also kept to this design, but later reduced the length and eventually introduced a short 'spike' bayonet.

When war broke out between Russia and Turkey in 1853 Britain and France joined the latter as allies in the fight against the former. The war became known as the Crimean after the area where the main operations were conducted on the Crimea Peninsula which extends into the Black Sea. The allies were also joined later by some Sardinian troops and other nationalities in a war which has come to be recognized by the names of so many famous battles and incidents such as Inkerman, Alma, Charge of the Light Brigade and the Assault on the Redan. By the time the war ended in February 1856 the casualty rate to all belligerents was enormous. The French sent 300,000 men to fight in the war and lost 11,000 killed in battle and almost twice that number through disease. The British army sent over 111,000 men losing almost 4,800 killed in battle and more than four times that number due to disease. The Turkish army lost 30,000 men and the Russians committed 500,000 men to the war and lost 110,000 due to all causes. The bayonet was used in many of the engagements, but the war would prove that the days of the bayonet charge were long over. Some of the senior officers had fought during the Napoleonic Wars, and the British army, whilst it had been engaged in many small wars around the British Empire, had not been involved in a war on this scale since the Battle of Waterloo. Their ideas, weapons and methods were old-fashioned and the Crimean War would have the effect of changing everything from tactics to the way in which troops were supplied and cared for on campaign.

On 20 September 1854 the British army found itself engaged in one of the earliest large-scale battles of the War. During the course of the battle the British soldier showed that he had not lost any of his fighting spirit and his ability to use the bayonet. As with all battles in this war it was a fierce engagement during which the French and British troops held the high ground and the Russians attacked uphill. The French and British lost 3,300 men and the Russians lost 5,700 killed. After the battle the journalist William Russell of the London Times newspaper singled out the achievements of the 93rd Highlanders for high praise and referred to them as the 'thin red streak tipped with a line of steel.' He was referring to the regiment's steadfastness in facing up to Russian cavalry charges with fixed bayonets in line as opposed to resorting to the more usual tactic of forming into squares, which was the traditional practice for infantry to repel cavalry. Over the years Russell's remarks have been turned into 'the thin red line'. The original remarks gave artists back in England the inspiration to paint the scene showing heroic Scottish Highland troops fighting against such adversity and adding yet another chapter to the legend of the bayonet.

Just over six weeks later the Battle of Inkerman on 5 November the British soldier was again using the bayonet in close-quarter fighting against a tenacious enemy. One of those involved in the fighting later wrote he: 'fell in with a small party who had taken cover under a low wall. I dropped to my knees and in a few moments a large body of Russians showed themselves, coming right straight for the wall. I fired we all fired on they came close to our barrier. We had not time to finish loading. Up we jumped, threw stones, bayonets, any thing we could lay ours hands, right at them. This we continued until quite a young officer with a Red jacket ran up seeing our fix and shouted, charge them lads. With this he jumped on the wall, we followed and let drive amongst them which laid them in heaps dead & wounded. Some of the wounded tried poor fellows to crawl away but was soon stopped & brought in. One man got up and was walking off quite well. I jumped over the wall again, behind which we had once more taken cover owing to our small number and took him by the collar and brought him in. His wound was a slight one.' This incident bears out the statements made by other witnesses and records of the Battle of Inkerman where they state the troops threw stones. Bayonets were also in use, certainly in this skirmish, and the charge was almost certainly made at the point of the bayonet with close-quarter fighting. Stone throwing was a natural reaction and done by all sides during the war. At the siege of Kars, June to

November 1855, the Turks were running short of ammunition and resorted to throwing stones at their Russian attackers who were advancing with fixed bayonets.

Assistant Surgeon John Scott was attached to the 57th Regiment of Foot (later to become the Middlesex Regiment) and he was able to make many observations, a number of which supported those of Sir Charles Bell and George Guthrie and even the French surgeon Larrey that very few wounds were caused by bayonets. On the 6 November 1854, the day after the Battle of Inkerman he noted in his diary that: 'medical duties were hard and dangerous yesterday, but today they are laborious in the extreme—amputations, extracting balls [bullets], dressing wounds &c. Our men are mostly wounded in the upper part of the body by gunshot. This probably arose from the close proximity of the enemy. We have only two or three bayonet wounds, and they were given after the men had fallen. The Russian prisoners however had many bayonet wounds, but indeed few compared with the number I expected. However, the enemy don't like too close quarters.' If one uses the calculation made by the historian William C. Davis estimates that only four men in 1,000 were killed by bayonet wounds during the American Civil War and applies it to the losses of the Crimean War the French lost only 44 men to bayonet thrusts. The British had fewer than 20 fatalities but the Russian and Turkish figures for losses to bayonets would produce a more ambiguous number because they record deaths due to all causes. Even if one applies the 2 per cent estimate as used at the Battle of Malplaquet from 1709, the number of deaths caused by bayonets to French and British would amount to 220 and 96 respectively. Thus, non-fatal wounds could be treated and, provided they were kept clean and free from infection, the wounded man would survive.

The Crimean War had been over for barely 15 months when Britain faced a crisis in the largest and richest territory in the Empire when rebellion erupted among the native military forces in India in May 1857. Rumour had been circulating that the new paper cartridges for the service rifle were greased with beef or pork fat and handling such was against the religious convictions of Hindu and Muslim troops respectively. The troops had to bite the end of the paper cartridge off before loading the weapon. The uprising was referred to as the Indian Mutiny but today is sometimes referred to as the Indian Rebellion of 1857. Indian troops rose up to protest against handling the new cartridges but a number did remain loyal to the British. As the unrest spread so the fighting spread with battles being fought as in any war and cities were besieged such as Delhi. The Mutineers captured the city in May and held it after initial fighting to seize it. British troops and loyal Indian troops moved in to lay siege to the city from June onwards and made preparations to storm the city. Finally, on 14 September an assault was mounted to recapture the city from the Mutineers and at the Kashmir Gate a number of Indian troops and British troops managed to fight their way through to engage the Mutineers at extreme close quarters where they used swords, bayonets and native daggers in the crush of battle. One British officer later recalled how troops attacking a large house defended by Indian troops: 'quickly made a lodgement on the ground floor and hunting the sowars from storey to storey at last bayonet met sword on the broad flat roof on which in a moment not a trooper remained alive'. Many were bayoneted but some were thrown bodily from the roof. In such circumstances it is difficult, if not almost impossible, for a soldier to suddenly stop fighting as the 'red mist' descends in front of his eyes. Under such conditions he will not accept surrender, and not until the last of enemy has been killed will a man cease, because there is no-one else left to fight. The city of Lucknow was besieged between 30 May and 27 November 1857 and before leading his men to attack the Mess House in the assault Sir Colin Campbell addressed his officers and emphasized to them: 'the necessity of using the bayonet as much as possible' and the men were not to halt to fire if they could avoid doing so.

The Indian troops who mutinied were brave men and had been trained in the tactics of the British army. They had the courage of their conviction and now used their training on the battlefield, which they turned against their British masters. In an assault to end the siege of Delhi, the 75th Regiment of Foot (later to become the Gordon Highlanders) were ordered to prepare to charge. On being given the command the men brought their muskets level to the position of 'engage' and continued to advance a few more paces as the line steadied and was observed: 'the line seemed to extend as each man sought more room for the play of the most terrible of all weapons in the hands of the British soldier… The long hoped-for time had come at last… and a wild shout or rather a yell of vengeance went up from the Line as it rushed to the charge. The enemy followed our movements, their bayonets were also lowered and their advance was as steady as they came on to meet us, but when that exultant shout went up they could not stand it, their line wavered and undulated, many began firing with their firelocks from their hips and at last as we were closing in on them the whole turned and ran for dear life followed by

a shout of derisive laughter from our fellows. In three minutes from the word
to charge, the 75th stood breathless but victors in the Enemy's battery.' The
Indians were very brave but against better-equipped, professionally trained
and properly led troops, they could not hope to hold any position.

The fighting continued across India and towns and cities held by rebel troops
were besieged and attacked to recapture them in turn, such as the walled town
or serai at Najufghur, which was assaulted in August 1857. Lieutenant Edward
Vibart, serving with the 1st Bengal European Fusiliers witnessed the attack
and recalled how: 'A column composed of ourselves, a wing of Her Majesty's
61st Foot and the 2nd Punjab Infantry, was then told off to attack it and, having
advanced to a point about three hundred yards from the building, we were
directed to deploy, halt, and lie down.' General Nicholson commanding and
other officers made an assessment of the situation and a battery of Royal Horse
Artillery came forward and opened fire on the position to give support to the
infantry attack. Lieutenant Vibart continues: 'The order was then given to the
attacking columns to stand up and, having fixed bayonets, the three regiments,
led by General Nicholson in person, steadily advanced to within about one
hundred yards of the enclosure, when the word of command rang out from
our commanding officer, Major Jacob, 'Prepare to charge!' 'Charge!' and in
less time than it takes to relate we had scaled the walls, carried the serai and
captured all the guns by which it was defended. Only a few of the rebels fought
with any pluck, and these were seen standing on the walls, loading and firing
with the greatest deliberation until we were close upon them. But few of these
escaped, as they were nearly all bayoneted within the enclosure.' This is another
account of the viciousness of close-quarter fighting with bayonets with neither
quarter being asked or given. The only way for rebels to survive the intensity
and ferocity of the fighting was to flee and hope they were not captured later.
Another witness to the scene later wrote an account how: 'Our guns went away
to the flank. We got "Fix bayonets, and trail arms; quick march!" On we went,
in a beautiful line, at a steady pace. On we went, and we got within some fifty
yards of them, when the men gave a howl, and on we dashed, and we were slap
into them before they had time to depress the guns. It was bayonet to bayonet
in a few moments, but we cut them up… We had very few men killed in the
charge, as we got in before they fired grape [shot]. Lieutenant G., 61st, was
bayonetted by a sepoy after cutting down two. N. shot the man that did it.'

The uprising lasted over a year and was eventually suppressed in June 1858
and the last of the Mutineers surrendered to face punishment, which was

often harsh and brutal. The situation was brought under control when the native troops were permitted to grease their own bullets using a compound prepared by themselves probably based on mutton fat. An alternative method of opening the cartridge paper instructed the troops to tear open the paper ends rather than biting. To troops used to biting actions for speed tearing by hand was awkward. Such a situation would have benefitted from Mr Richardson's serrated device fitted to the bend to open the cartridges, but by this time it had been forgotten about. Bayonets had been used by both sides during many of these clashes between; but even so, it would appear that the number of bayonet wounds recorded was very low. Following one engagement a study of 60 wounded British troops reveals that only two were caused by bayonets, which represents 0.33 per cent of the total number wounded, while 12 were wounded by sword and lance which represents 1.99 per cent of the total. The remainder were caused by firearms, some of which were ageing in design. The Indian Mutiny was not the only unrest the British army had to face as it 'policed' Britain's Empire and other European states faced similar unrest in their overseas territories. France too had its own troubles in Mexico and North Africa and in most of these 'little wars' the bayonet was used to put down insurrection along with better firepower and discipline.

By the time the American Civil War broke out in 1861 pamphlets and training manuals were being used to instruct new recruits in weapon drill and firing and this included the use of the bayonet. For example, the manual entitled 'Skirmishers' Drill and Bayonet Exercise' which was used by the French Army at the time and translated for the American forces by Lieutenant Colonel R. Milton Cary. It contained all the movements of the time used to counter most situations, including being attacked by cavalry. The instructions given for this action recommend that: 'The infantry soldier who is a good shot, and, at the same time, a good bayonetman, waits standing fast, for the horseman who charges him.' The manual goes on to instruct that an infantryman should not fire until the rider is as close as six to eight yards. Then he would be in a position to receive the cavalry armed with either sabre or lance. The manual comments that: 'The bayonet is the weapon of the brave' and it would have taken brave men indeed to face up to a cavalry charge in a modern battle during the American Civil War.

The war was fought over the vast country of America, which dwarfed Europe in size, but it would also prove to be the swansong for bayonet charges as artillery came to be the dominant weapon on the battlefield. At

Ephraim Graham Co B 45th Illinois Infantry. He is carrying the Yataghan bayonet.

some battles, such as Gettysburg on 1–3 July 1863, for example, where a total of 657 pieces of artillery was deployed, the figure represented more artillery than had been deployed at the Battle of Waterloo, where the number of guns was 455 between the French and Allies. Artillery could wreak terrible havoc among the infantry as it advanced, as happened at the Battle of Fredericksburg between 11 and 15 December 1862. A Confederate officer later wrote of the battle how he saw: 'The enemy, having deployed, now showed himself above the crest of the ridge and advanced in columns of brigades, and at once our guns began their deadly work with shell and solid shot. How beautifully they came on. Their bright bayonets glistering in the sunlight made the line look like a huge serpent of blue and steel. The very force of their onset levelled the broad fences bounding the small fields and gardens that interspersed the plain. We could see our shells bursting in their ranks, making great gaps; but on they came, as though they would go straight through and over us. Now we gave them canister, and that staggered them. A few more paces onward and the Georgians in the road below us rose up, and, glancing an instant along their rifle barrels, let loose a storm of lead into the faces of the advance brigade. This was too much; the column hesitated, and then turning, took refuge behind the bank.' Canister fired from artillery was like giant shotgun cartridges full of either musket balls or larger pellets, which spread out on being fired. At close range this type of ammunition was lethal and cuts swathes through the ranks of infantry in the same way that the larger grapeshot did against men and horses.

Armies in this conflict which saw neighbour pitted against neighbour and brother against brother numbered in the tens of thousands and the casualty rate was accordingly high. At the four-day battle of Murfreesboro the Union forces deployed 43,400 men against 37,700 Confederate troops and the two sides lost 13,249 and 10,266 killed and wounded respectively. At this time it was estimated that between four and six per cent of casualties were caused by bayonets and, taking this as an average, the combined casualty list from Murfreesboro gives us a figure of between 940 and 1,411 caused by bayonets. The historian William C. Davis calculates that around 0.4 per cent of all deaths in the war can be attributed to bayonets and swords. This is around four deaths in every thousand. Most sources state that around 620,000 men died in the war with the Northern Union forces losing 360,000 and the Southern Confederate armies losing 260,000. Of this figure it is believed that some 414,000 men died from disease and other injuries not related to battlefield wounds. This reduces the overall figure of those killed on the battlefield as around 206,000 for all causes. If one applies Davis' calculation to this figure

Recreated Confederate troops with socket bayonets fitted to muskets.

we arrive at just over 800 deaths due to sword or bayonet. The Union armies and Confederate armies recorded almost 282,000 and 194,000 non-fatal wounds respectively. Using the same calculation to these statistics the Union had 1,124 sword and bayonet wounds and the Confederate forces suffered 776 wounds of the same kind.

By the time the war ended in 1865 more than 2.7 million men had served in the ranks of the Union Army of the Northern States and the Confederate Army of the Southern States had some 750,000 enlistments. The north was heavily industrialized and during the war produced its own artillery and over 2.5 million muskets. The south had little in the way of industry and imported 600,000 weapons from Europe and Britain and the troops used whatever they could collect from the battlefields. It was really only in this way they could support the armies in the field. For example, on 14 June 1863 following an engagement at Winchester, the Confederates captured 200,000 rounds of ammunition which would have only been sufficient to fight part of an engagement at least. During the early months of the war the disparity in calibres with so many different types of firearms was a logistical nightmare for both sides. This variety of weaponry extended to the types of bayonets, also. By and large, however, it was the standard socket-type with its long pointed triangular-section blade that was the most widely used type. This style could be up to 18 inches in length but some such as the bayonet designed for use with the Brunswick rifle used by the British army were 22 inches in length and weighed 2lb. Another style of bayonet was the Yataghan, which had a curved flat blade. It was popular in France, where it had been adapted, and was of Turkish influence. Other armies, including the British army, would use the style for a period and sometimes referred to the style as the 'lunger'. The Brazilian army would be one of the last armed forces to use the type when it accepted it for service in 1904.

Some troops serving in the war were immigrants from Europe and had served in wars involving countries such as France, Prussia, England, Austria and any of a dozen other states. These men were experienced in warfare but American-born recruits had to learn their lessons of war on the battlefield. A few senior officers had been to Europe to observe wars but this was not the same as being involved directly with commanding a battle. The result was some very old-fashioned ideas on how a battle should be conducted. The theory such as firing two or three volleys then advancing with bayonets levelled was fast becoming an anachronism and when attempted against artillery the outcome was costly in manpower. At the Battle of Bull Run 21 July 1861, Confederate General Thomas J. Jackson turned to an officer and declared about the Union troops advancing: 'Sir, we'll give them the bayonet.' He had seen action against Mexico between 1846 and 1848, but some of his ideas were outdated. He exhorted his men to: 'Trust to the bayonet.' The outcome of the battle resulted in a victory for the Confederates who lost almost 2,000 killed and wounded out of a force of 32,500 troops. The Union Army had lost almost 3,000 killed and wounded out of a force of 35,000 men.

Union soldier with socket bayonet and triangular blade fitted to musket.

The term bayonet charge was being used but during the war but as it progressed the term and the very act itself became used less and less. At the Battle of Fort Donelson in Tennessee on 11–16 February 1862, a Confederate colonel who observed an attack by Union forces remembered that: 'This was not, strictly speaking, a 'charge bayonets', but it would have been one if the enemy had not fled'. European countries sent observers to see for themselves how the war was being fought and report their findings. Communications with telegraph and faster ships crossing the Atlantic meant that news of this war could reach Europe in days where developments were read by more military commanders. They believed that lessons could be learned from this war and tactics had to be changed. Some officers realized the day of the bayonet charge was over, but others stubbornly adhered to its outmoded role and maintained a bayonet charge could still turn the course of a battle. Developments in weaponry would show who was right and who was wrong and this was evident in Prussia where armaments manufacturers such as Alfred Krupp were producing artillery of a powerful nature that would change the nature of warfare. In France the early machine guns in the form of the multi-barrelled Mitrailleuse were being developed and breech-loading rifles increased an infantryman's firepower. During the American Civil War the Union troops began to use the Henry rifle, which could be loaded with seven rounds of ammunition and greatly increased the rate of fire. The Confederate forces who came up against it cursed the weapon, claiming: 'that damn Yankee gun that can be loaded on Sunday and fired all week.' Such innovations in weaponry were being appreciated by the armies in Europe, but they were also learning their own lessons in modern industrialized warfare.

Less than ten years after the conclusion of the American Civil War officers who had seen service in conflict, including General William T. Sherman, were suggesting that the bayonet no longer had any practical use on the battlefield and that it should be replaced by something more useful. This opinion may have stemmed from the fact that so few battle casualties could be attributed to its use in battle. Sherman did not continue by making any suggestions as to what implement could be used to replace the bayonet, but there were a number of officers who disagreed with him and who felt that if the bayonet was taken away from the infantry they would be denied half the capacity to fight. That may seem like an over-exaggeration but the sentiments explained the way the infantry felt about its bayonets. One idea to replace the traditional bayonet could have been the so-called 'Trowel Bayonet' developed around 1870 by Lieutenant Colonel Edmund Rice, who had served in the Union Army Civil War seeing action at Gettysburg and Spotsylvania Court House where he was wounded.

After the Civil War Rice served in the Indian Wars and was very active in many field of interest and he devised the idea of the Trowel bayonet and took out the US Patent 91,564 in June 1869. Eventually some 10,000 produced and a number were used on campaign against the Nez Perce native Indians in 1877 and were fitted to the Model 1873 Springfield rifle. The Trowel bayonet was even demonstrated to European armies but it was not taken into service. It did attract some attention in America where it was seen as being an entrenching tool, but beyond that very little interest was given to the design and the whole concept was declared obsolete in 1881. The French had developed a similar style around 70 years earlier but that design did not enter wide service. The Rice design was a socket type measuring 14.25 inches in length overall with the blade being 10 inches in length and 3.5 inches across at its widest point. The term came from the fact that the flat triangular blade resembled a bricklayer's trowel and despite its size it weighed less than 1lb, but the wide shape of the blade would have produced terrible tearing wounds. Bayonets may have been used as a multi-purpose tool for chopping wood, cooking, candle holders and even tent pegs, but each man felt he had something handy he could rely on especially the flat, broad-bladed Yataghan bayonets. Admittedly, there was little if any need of use of the bayonet in the wars against the Native American Indian tribes between 1865 and 1898, due to the style of fighting which was largely guerrilla tactics on the part of the natives who did not fight in the conventional manner of formalized battles. American troops did use the bayonet to limited degrees in the Spanish-American War of 1898 and the fighting on the island of Cuba and later during operations along the Mexican border until 1917 when America declared war on Germany. In Europe there was never any question about the future of the bayonet. In the second half of the nineteenth century it was not as significant as it had once been, but armies engaged in European wars and overseas campaigns still had a need for it.

The many wars of the time saw armies of increasing size being mobilized and railways were speeding up the process of moving troops over vast distances. The Austro-Prussian War of 1866 saw troops using bayonets in battles such as Sadowa, during the Russo-Turkish War of 1877–1878 both sides used bayonets in battle and during the Franco-Prussian War of 1870–1871 both the French Army and the Prussian Army used bayonets, although its importance on the battle was gradually becoming less with each passing war the bayonet was still capable of being used. Across the great continent of Africa European troops were being used to control the overseas territories of nations such as Belgium in the Congo region, France in several locations where it would recruit local natives to raise regiments of Zouaves, Turcos, Spahis and Chasseurs d'Afrique, all of which took to the use of bayonet very willingly. The recently united Germany became involved in Togoland from 1884 and the Britain which had the greatest interest and financial invest of all European nations built railways and expanded the Empire. The British army maintained these territories and faced many different native groups from the vastly numerical such as the Zulu, the fanatical Dervishes and the Boer settlers who fought for independence, but the prospect of gold, diamonds and other riches was too great and in the case of the latter enemy the British army faced a determined force of disparate farmers in a bloody and protracted war. Again, as with the Indian War in North America, conditions and guerrilla tactics precluded the possibility of any formal set-piece battle and sieges at locations such as Ladysmith developed. In these situations there was little use for the bayonet except to guard prisoners kept in stockades, the same way as sentries during the American Civil War had guarded their charges at prisoner of war camps.

The decade of the 1860s was a turning point in weapon design for several European armies, including the British army, during which time they took into service new types of rifles. Prussia and France introduced bolt-action, breech-loading rifles Zundnadelgewehr 'Needle Gun' and the 11mm calibre 'Chassepot' Modèle 1866 respectively, and the British army joined this arms race by introducing the Snider-Enfield rifle in 1866. This was the first breech-loading rifle for the British army and kept it in line with the other European armies and their breech-loading rifles. The Snider weighed 8lb 14oz and was designed by the American-born Jacob Snider, and fired a rimmed cartridge of .577 inch calibre with the bullet head of 31.056 grains reaching a muzzle velocity of 1,240 feet per second with an effective range of 300 yards. Each of these weapons was equipped with its own type of bayonet and with muzzle-loading a thing of the past infantrymen could now operate in a more flexible manner on the battlefield. For example, the standard bayonet for use with the Chassepot rifle measured 27.5 inches in length and weighed over 1lb 11oz. Its long recurved blade with fullered grooves along the length was a style known as the 'yataghan-type' that was used in America and later the British army developed this style of bayonet for use with the Martini-Henry rifle and was sometimes known as the 'lunger' bayonet. The French bayonet was fitted with a spring-loaded locking bolt on the handle to attach it to the rifle a muzzle ring was incorporated in one of the quillons, which was also fitted with a locking screw to tighten it to provide a secure fit to the rifle. The Chassepot bayonet had the lower point of the other quillon curved forward in a hook shape which gave the bayonet a distinctive appearance and this too was also copied by other countries such as Japan. The hook was called a blade-breaker, but given the fact that bayonets are made of steel it was therefore almost impossible to have been used to break an opponent's bayonet blade. Instead the best the shape could have been used for was to have hooked the blade of an opponent in order to deflect the bayonet thrust.

The Snider had only been in service for six years when a new rifle was introduced for service and this was the Martini-Henry. The new design was a combination of a barrel designed by a Scottish gunsmith, Alexander Henry, and the action designed by an Austrian, Friedrich von Martini. The first model to appear was known as the Mk.1 Martini-Henry in .450-inch calibre and was first issued to British troops in 1871before entering general service by

1874 at a time when the strength of the British army stood at around 190,000 men. It was not universally liked by the troops and developed a reputation for having a fierce recoil action on being fired. In his work 'The Book of the Rifle' published in 1901, writing as T.F. Fremantle, Lord Cottesloe wrote of the recoil action of the Martini-Henry as: 'The 'kick' of this rifle was a terror to the unfortunate recruit who for the first time experienced its violence. There were few men who did not find a comparatively small number of shots fired during the day were enough to take the edge off the accuracy of their shooting. Many were the bruised shoulders for which the rifle was responsible.' Despite this the weapon was undeniably accurate out to extreme ranges and although sighted to 1,000 yards the Martini-Henry rifle gained a reputation for being capable of making shots out to extreme ranges. General Sir Arthur Cunyngham KCB, recorded such an incident during the Kaffir Campaign of 1877–1878. He wrote how: 'All of them [rifles] were eclipsed at the Waterkloopf when the Sergeant-Instructor of Musketry of the 90th Perthshire Light Infantry killed a Kaffir by deliberate aim at 1,800 yards distance – a little over a mile!... one of the enemy made himself defiantly conspicuous to a party of the 2nd Battalion 24th Regiment. Several shots were fired at him, which caused the fellow gradually to increase his distance. At slightly over 1,000 yards the native appeared to consider himself safe; but an officer came on the scene, and at his first shot the whooping and dancing Kaffir received a fatal bullet between the shoulders.'

The Martini-Henry was used in many of these frequently bloody campaigns, which were often concluded without protracted warring due to the deployment of superior European weaponry against which local natives did not stand a chance, despite their overwhelming numerical superiority. The Martini-Henry rifle was four feet and 1.5inches in length and capable of being fitted with a socket-type bayonet known as the Type '73 after the year of its year of introduction. The blade was the standard triangular form in section measuring 22 inches in length and although an old design it was this type of bayonet that troops used in many close-quarter fighting actions against the likes of Zulu warriors where its reach of six feet with a thrusting action would have been formidable. A version of the rifle was also used by the Turkish army, but this was a combination of the Martini and Peabody designs and was actually known as the Peabody-Martini rifle. In 1877 the Turkish army using these weapons inflicted 37,000 casualties on the Russian army at the Battle of Plevna in July that year.

Recreated scene showing bayonets used against a Zulu warrior.

The British army faced its greatest adversary in 1879 when events in South Africa led to the outbreak of the Anglo-Zulu war in January that year. The Zulu and British armies fought a number of fierce battles over a period of several months, but the Zulu War has come to be epitomized by the two main opening engagements fought at Isandhlwana and Rorke's Drift, which were fought within hours of one another at two separate locations. The first action came on 22 January 1879 when a British force comprising of six companies of 24th Regiment of Foot were attacked at Isandhlwana by a Zulu army of at least 14,000 warriors commanded by Matyana. Using mainly short stabbing spears, called assegais, this Zulu force killed over 1,500 men who had been armed mainly with Martini-Henry rifles. It was the worst disaster inflicted on the British army by any native force and inspecting the remains of the battlefield showed how some troops had fought bayonet against assegais and some men had formed themselves back to back to fight using the bayonet until worn down and killed where they stood.

Immediately after this action a force of between 3,000 and 4,000 Zulus made their way towards Rorke's Drift, a mission station which was being defended by 140 men of the 24th Regiment of Foot commanded by Lt. Bromhead

British troops at the Battle of Tel El-Kebir, 13 September 1882.

Informal image of unit of British army in camp. The men at either end carry Martini-Henry rifles with triangular-bladed bayonets.

of the 24th Foot and Lt. Chard of the Royal Engineers. The defenders at Rorke's Drift fought off repeated attacks by the Zulus and it is estimated they fired some 20,000 rounds of ammunition and killed at least 400 Zulus for the cost to themselves of 25 killed and wounded. The close-quarter fighting was intense and during the course of the action at Rorke's Drift on 22–23 January which lasted approximately twelve hours the defenders made extensive use of their bayonets as they crossed blades against the attacking warriors armed with short stabbing spears known as assegais. In the hand-to-hand fighting the Zulus grabbed at the bayonets trying to wrench them off the soldiers' rifles and in a few instances they actually succeeded. Fighting with weapons such as the bayonet against the spear was reminiscent of warfare 500 years earlier during the medieval period, but here in the late nineteenth century it was still proving an effective way of fighting even though it was born out of desperation. During such hand-to-hand fighting men could not hesitate or feel any repulsion about stabbing a man completely through his abdomen. Squeamishness was something the men of the 24th Regiment of Foot could not afford as they defended their hastily barricaded position and as the fighting intensified they became like automatons thrusting at the glistening bodies of the Zulu warriors as they clambered over the barricades and forced their way into the interior of the defences. It was all a matter of survival instinct and the bayonet practises instilled into each man at the training depot must have returned as natural reflexes as they lunged forward to meet each charge of the Zulu masses. After the battle eleven of the defenders were awarded the Victoria Cross, the highest decoration for bravery in the British Army, one of whom was Corporal Christian Ferdinand Schiess of the Natal Native Contingent. A description of his actions recalled how a Zulu warrior armed with a Martini-Henry rifle fired at Schiess who despite his wounds: 'jumped on to the parapet, bayoneted the Zulu, regained his place, bayoneted another, and then climbed once more upon the sacks and bayoneted a third'. It describes the intensity and fierceness of the battle as the defenders struggled to hold their position.

A veteran of the action at Rorke's Drift later recounted how to deal with the Zulu warriors at close quarters the troops needed 'the old three-sided bayonet and the long thin blade we called the "lunger".' He observed these were 'fine' weapons but he noted how some were 'very poor in quality, and either twisted or bent badly. Several were like that at the end of the fight; but some terrible thrusts were given, and I saw dead Zulus who had been pinned to the ground by the bayonets going through them.' The Zulus finally withdrew in the face of the stoicism and controlled fire of the British soldier with his Martini–Henry rifle. The war continued until August that year when Cetewayo was captured. The cost to the Zulu army was enormous and a true casualty figure will probably never be known. What the Zulus lacked in equipment they made up for with a huge reserve of manpower which some sources calculate had a strength of 50,000 warriors that allowed it to take on Britain which was the most powerful nation in the world at that time. Britain accepted that a war against the Zulu nation had been inevitable but the military commanders did not expect to be attacked by such a fierce and highly disciplined force. This war taught many armies around the world a valuable lesson, and the British soldier added the Zulus to the list of those whose bravery in battle they respected. Exactly how many Zulus were killed by bayonet thrusts alone is not known and those killed by a combination of bayonets and gunshot wounds is not known either, but it would be the superior firepower of machine guns in later engagements and other campaigns which would prove the deciding element. The bayonet was there to back up such weaponry if the enemy did manage to get too close.

The Martini–Henry rifle was used in many of the major campaigns and some lesser engagements of the time in which the British army saw action across the vastness of the Empire, including Canada and South Africa. At the Battle of Tel El-Kebir on 13 September 1882 British troops fixed bayonets for charging and also for defence against fierce attacks. After the battle General Sir Archibald Alison remarked that he had: 'never seen men fight more steadily. Five or six times we had to close with them with the bayonet, and I saw those poor men fighting hard when their officers were flying before us. All this time, too, it was a goodly sight to see the Cameron and Gordon Highlanders mingled together as they were in the stream of the fight, their young officers leading in front, waving their swords above their heads, and the men rushing on with that proud smile on their lips which you never see in soldiers save in the moment of successful battle.' This observation is full of jingoism, which the British public wanted to read in their newspapers. The smile on the men's faces was probably due to nerves, as no man wants to let down his comrades in battle by showing concern. This imagery of men smiling as they went into battle was portrayed in many paintings by artists such as Lady Butler who depicts men smiling as they form square to counter cavalry charges at the Battle of Waterloo. Other artists for popular journals at the time also portrayed men smiling as they charged or stood up to the enemy at engagements such as Rorke's Drift and Isandhlwana and this was repeated in other countries such as France where troops were engaged in campaigns against natives in their overseas territories. In 1909 American forces were involved in dealing with insurrection outside the mainland in overseas territories in the Philippines. The troops were armed with the Springfield M1881 rifle fitted with bayonets but they also used shotguns for close-quarter action. These weapons such as the Winchester 12-gauge pump-action were fitted with a bar to accept a bayonet. At one engagement on the island of Sulu a group of Moro rebels were surrounded and during the engagement the fighting developed into a: 'bayonet stabbing melee… during which Baer killed three more Moros…while a sailor skewered a fourth with his bayonet.' The fitting of bayonets to non-standard weapons such as shotguns would continue and when America entered the First World War in 1917 shotguns such as Remington M10 'Trench Guns' were fitted with adaptors to accept the M1917 rifle bayonet or the Russian Mosin-Nagant bayonets.

During the course of its service life the Martini-Henry rifle underwent a series of modifications to produce four different versions, known as marks, and bayonets were also produced or adapted for use with these variants. In 1875, units of the Royal Artillery were issued with the Martini-Henry rifle for which they were issued a bayonet with a blade length of 18 inches, of which 8 inches on the back edge was specially ground to produce a cross-cut saw pattern. In 1879 the blade length was increased to 25.75 inches, which was a mighty blade and not far off being a sword in its own right. This was followed up seven years later in 1886 and again the following year in 1887 with new pattern sword-bayonets, with blade lengths of 18.5 inches, and were introduced for the line regiments. The Pioneer's sword-bayonet of 1856 was modified and trials were carried out to fit it to the Martini-Henry rifle were made in 1895 and it was also tried on the later rifles such as the Lee-Metford and Lee-Enfield but without much success because in the opinion of some of those of the day it was being used for too many things beyond its original design and would simply not work.

By now some bayonets with blades of quite exceptional lengths were being produced for armies around the world and were known as sword-bayonets. The style actually dated back to the turn of eighteenth and nineteenth century with one of the earliest forms to be used being the P1788/1801 for the Danish army and indeed looked like a sword including the knuckle guard on the handle. It was fitted with a bar that faced towards the pommel to attach it to the musket. Another sword-bayonet design, which appeared in 1810, was developed by the London-based gunmaker Staudenmayer. This was produced for volunteer troops serving in England as opposed to serving overseas in Spain. This bayonet weighed 17.5oz and measured 30.5 inches in length. The muzzle of the barrel passed through the brass knuckle bow and locked on the quillon to provide a firm fixture and lock the bayonet in place. Volunteer units served in Britain for home defence duties during the Napoleonic Wars and some of these were issued with sword bayonets. There were a few units that were equipped privately from funds of local businessmen or officers and there were designs, which used a sword-type handle to clip into the standard socket bayonet to turn it into a sword bayonet.

Over the next few years, the style was copied and extended from use by infantry units and issued to specialist types of troops, such as pioneers who could use them as a tool as part of their engineering role. Sword bayonets of the Pattern 1841, which incorporated a serrated or 'saw' back to the blade, were issued to sappers and miners who were armed with carbines. Between 1845 and 1854 the sappers of the East India Company were issued with sword bayonets for use with their carbines and later, even the American military followed the trend and introduced the type into service. Committees were established to look into the usefulness of the design and a popular journal of the day carried its own opinion of the bayonet and commented: 'In recommending this new sword bayonet, they appear to have had in view the fact that bayonets will henceforth be less frequently used than in former times as a weapon of offence and defence; they [the committee] desired, therefore, substitute an instrument of more general utility'. Here too there were calls to replace the bayonet with an all-purpose tool but it was to prove a stubborn weapon to replace. Soldiers by habit would rather utilize what they already have than have something new in place of a reliable item which in this case was the bayonet.

The bayonet is traditionally seen as being primarily an infantry weapon and as such used in land battles. When marine units were created and deployed to serve as guards on warships armed with muskets it was only natural that they also be equipped with bayonets. These proved to be very handy weapons during the early days of sailing ships which were armed with muzzle-loading cannon when ships may come alongside as crews tried to board each other's vessels and fighting at close quarters was expected. Marines could fire their muskets and then use their bayonets in the same manner as the infantry on land, but being fewer in number they could never form together to present a line of bayonets. Furthermore, the space on board a ship was very limited and movement severely restricted. This meant marines had to fight either as individuals or perhaps two or three men together. They were in effect left to use their own judgement when it came to fighting such boarding actions. Later on naval ratings were used to send landing parties ashore to support operations and for sentry duties they were issued with service weapons and bayonets. In 1858 the British Royal Navy began using the naval cutlass bayonet, which weighed 2lb 6oz and measured 25.5 inches in length and used on the Enfield Short Naval Rifle. The Royal Marine Artillery also used the same style of bayonet. Some 30,000 of these bayonets were produced by bladesmiths in the Belgian town of Liège and a further 418,000 were produced in the English city of Birmingham, which was best known for its textile industry, but as a manufacturing base was sufficiently adaptable to manufacture steel blades for bayonets. The numbers of these naval cutlasses produced far exceeded the numbers required by the Royal Navy and these excess bayonets were sold to overseas navies. In 1871 another style of bayonet was introduced for use with the Martini-Henry rifle and this weighed 1lb 15oz and measured 22 inches in length. The long lengths of the blades of these bayonets made them clumsy but that did not prevent other navies from following suit. For example, the Navy followed the trend with the M1870 sword-bayonet with a long flat blade with a spring-loaded catch in the handle and a muzzle-ring in one of the quillons.

Pioneer troops serving with engineer units were issued with brass-hilted sword-bayonets, which had a serrated or saw-backed edge, and the blade was stout enough to be used to saw wood and break open boxes containing supplies. These designs resembled tools intended for heavy-duty work rather than the refined shape intended for thrusting into the body of an opponent. The British army used this style and pioneers in the Russian army in the 1840s were issued with a heavy-bladed, saw-backed sword-bayonet, also. This design measured 25 inches in length and weighed 2lb 8oz. It was based on the design used by the French army in the 1830s, which in turn resembled the ancient Roman sword Gladius Hispaniensus from 2,000 years earlier.

There is no questioning that the bayonet was useful in defence against cavalry or the onrush of massed infantry, especially when used en masse in forming square to present a hedge of steel points which no self-respecting cavalryman or infantry would dare attempt to assail or even venture near. The first design of bayonet fitted for the Martini-Henry was a modified type originally developed for the Snider. This was done by 'bushing out' the socket ring by fitting a reducing ring so that it fitted the muzzle of the barrel and this remained the standard issue for the time being, with a blade length of 17 inches, until a new bayonet was introduced specifically for the Martini-Henry in 1876. The socket ring attachment was retained but the blade length was increased to 21.5 inches with a triangular shape in cross-section. The troops called it the 'lunger' and it did give the infantry a slight edge in length of reach when thrusting or lunging. The year before its introduction the members of the Royal Artillery had been issued with their own unique design of bayonet for the Martini-Henry and this was a straight-bladed weapon with a blade length of 18 inches into which a cross-cut saw pattern was cut for about 8 inches along the back of the blade. Five years later in 1879 the length of the blade was increased to a fearsome 25.75 inches, another 4.25 inches greater than that of the standard infantry bayonet for the Martini-Henry.

In 1886 the French army accepted into service the Lebel rifle that was equipped with the Épée-Baïonette Modèle 1886, which had a blade measuring 20 inches in length with a square cross-section which gave it a spike-like appearance. When fitted it gave the infantrymen a lunging reach of six feet and was the style which would see it through many Colonial wars and the First World War. In the same year the British Army accepted new pattern sword-bayonets with blades measuring 18.5 inches in length for the Martini-Henry and for some inexplicable reason the following year another new bayonet almost identical was introduced for the same weapon. It was during the campaigns against the Zulus and other native tribes such as the Dervishes in Africa that the quality of bayonets used by the British Army began to be questioned because so many were distorting and twisting out of true line to such a degree that in many cases a second thrust was not possible. The quality of the bayonets was questioned in the newspapers and in the House of Commons the Government found itself having to explain matters regarding the deficiency of a very basic weapon.

The matter soon became referred to as the 'Bayonet Scandal' when newspaper journalists such as Burleigh who wrote for the *Daily Telegraph* published their reports. In the case of Burleigh he had been a witness to the fighting at El Teb and Abu Klea and the advance to Khartoum and seen for himself the very nature of the fighting. In one of his reports he wrote an account of how he had personally seen 'Many a soldier at Abou Klea saw with dismay his bayonet rendered useless at the moment when there was no chance to load his rifle, and when he most stood in need of its services. There also I saw sword-bayonets bend and twist with the facility of soft iron rather than steel.' In the aftermath of such damning evidence and criticism a series of tests were conducted and during one set of trials 50 blades broke, 300 were found soft and 750 were declared 'bad'. The same was found throughout the rest of the bayonet stocks in the army. The question was, how had the quality of bayonets been allowed to deteriorate to such a degree? Even soldiers and officers were asking questions about the quality of their bayonets. The problem stemmed from trying to cut costs and bayonets were purchased from contractors in Germany. They supplied sub-standard bayonets, which were then issued without the proper quality control checks. Not until their actual use in battle did the problem become manifest, by which time it was too late for many soldiers. It was decided to cease procuring bayonets from sources in Germany and production returned to English companies such as Wilkinson and Mole, who became responsible for the manufacture of all bayonets. They used the highest quality material, which was hardened, and eventually these moves, along with strict quality control and regular tests, resolved the crisis.

As the nineteenth century drew to a close several European countries were engaged in conflicts in their overseas colonies. France, for example, was maintaining Algeria in North Africa and the British Army was engaged in the Second Boer War in South Africa, which would continue to 1902. During this war the British soldiers had little opportunity to use bayonets in charging and instead it was used mainly fixed to rifles to by the sentries guarding the Boer prisoners held compounds. America had engaged Spanish forces on the island of Cuba during the Spanish-American war, which was concluded in just over three months, April to August 1898, and led to the collapse of the Spanish Empire. News of this event rocked Europe, where military commanders believed they were the decision-makers in the world. Japan was still an emergent nation at this time but its victory over the Russian Army at Port Arthur in 1904 during the Russo-Japanese war of February 1904 to September, and subsequent overall victory during the war elevated its global status. The war was termed as 'the first great war of the 20th century'. The Russians were armed with the new Mosin-Nagant rifle and the Japanese also used bayonets during massed, headlong charges against prepared Russian positions. It has been estimated that some 7 per cent of all casualties during the war were caused by bayonets. A post-war study of the Russian casualty figures was completed by Colonel Louis A. La Garde of the US Army Medical Corps, who made life-long study of battlefield wounds and between 1876 and 1877 had served during the 'Indian Wars'. Colonel La Garde discovered that out of 170,600 casualties he was only able to find 680 instances of bayonets being the cause of the wound. This was far fewer than the estimated 7 per cent figure and indicates that modern machine guns, artillery and better rifles were the main cause of deaths and wounds.

At this time there was a move in the US Army to introduce so-called 'ramrod' bayonets to fit the M1884 Springfield rifle. This rod-like device, resembling a ramrod, extended to a length of 35.5 inches, but the American President Theodore Roosevelt would have none of it and believed that the 'ramrod bayonet is about as poor an invention I have ever saw.' In correspondence he wrote of the bayonet that it: 'broke off short as soon as hit with even moderate violence. It would have no moral effect and mighty little physical effect.' The idea was dropped and joined the list of bad ideas for bayonet designs. At the beginning of the twentieth century there were questions being asked concerning the need for bayonets at all during any future conflict and many believed it was unnecessary and that it served no practical purpose in modern wars. It was not the first time the continued use of the bayonet had been questioned, but with so many innovations in weaponry it did seem that the bayonet might be nearing the end of its usefulness on the battlefield. Nevertheless, in training depots around the world soldiers continued to be instructed in the art of bayonet fighting. It was a training which would come to stand many of them in good stead during the war which would engulf most of Europe and eventually draw in countries from around the world and prove that the era of the bayonet was far from over.

Chapter 8

Bayonet Practice

Using a bayonet requires a man to be filled with aggression towards his enemy who he must be instructed to hate so that if he should confront him face to face he will not flinch. It sounds frightening and in the melee of battle men often found themselves up close and personal against an opponent who had also been instructed in a similar manner. Bayonet fighting also required physical strength and things happened very fast that it left little time to think. Men reacted by instinct and used anything immediately at hand to fight an enemy. This could be stones, lumps of wood and the bayonet, hand axes or daggers. It was very primitive to fight in such a manner especially in the late nineteenth century when weapons were becoming ever more powerful. The musket and then rifle was the infantryman's main weapon and he had to be taught how to load, take aim at a target and shoot with a degree of accuracy. This could be difficult and when the additional weight of a bayonet was added to the muzzle a man had to be taught how to control the drop and compensate for this. The early bayonets were heavy cumbersome things but a man could be taught how to become used to this by drilling with the bayonet fixed at all times. The Prussian army of the eighteenth century used this technique and by the nineteenth century the practice had spread. By using muskets with bayonets always fitted, infantrymen learned how to adjust their aim. Even with modern weapons fitted with lighter bayonets an infantryman has to take the attached bayonet into consideration.

New recruits joining the army have to be taught many skills, including marching and how to use all weapons in service and this includes bayonet practice. This part of a soldier's training has traditionally involved fitting the bayonet to his service rifle and charging at a sack filled with the straw suspended from a wooden frame. During his service in the Grenadier Guards in the 1970s the author of this book underwent this instruction during basic training at the Guards Depot at Pirbright in Surrey and remembers thinking even then how old-fashioned it was to be taught something which belonged on the battlefield of Napoleon and Wellington. Even the future Field Marshal

Sir Bernard Montgomery received such instruction as he mentioned in his Memoirs in 1958 which he recalled as: 'In my training as a young officer I had received much instruction in how to kill my enemy with a bayonet fixed to a rifle. I knew all about the various movements- right parry, left parry, forward lunge. I had been taught how to put the left foot on the corpse and extract the bayonet, giving at the same time a loud grunt. Indeed, I had been considered good on the bayonet-fighting course against sacks filled with straw, and had won prizes in man-to-man contests in the gymnasium.' He continued in his reflections how as a young Lieutenant serving in France during October 1914 he was confronted by a 'large German', but he was unarmed apart from his sword, without thinking he leapt at his adversary and kicked him very hard between the legs, which rendered him helpless, and Montgomery took his first prisoner. It seems remarkable that after so much training and skills in weaponry he should resort to the most basic instinct and kick his enemy.

Bayonet practice is often taught on individual level with each man taking his turn to run towards a straw-filled sack target. In previous wars recruits were taught to fix bayonets and advance en masse. Even so, a soldier will sometimes revert to type even without giving his actions a second thought and use the butt of his rifle as a club. Prince Frederick Charles of Prussia was watching bayonet practice when he saw some of the men reverse the rifles and use them in a clubbing action against the targets rather than stabbing at them with the bayonet. He approached one of the soldiers and asked why he did this this. The soldier allegedly replied he did not know except that: 'When you get your dander up the thing turns round in your hand of itself'. Certainly the clubbing action was nothing new, musketeers had done it in battle during the seventeenth century, and in battle some things do indeed happen automatically for which there is no explanation. During the Napoleonic Wars a French recruit underwent two months training during which time he would be taught bayonet drill either as an individual or en masse with hundreds of other troops to assault a position. Even after such instruction he was still just as likely to use his musket as a club in battle as a last resort.

This same kind of training in bayonet drill was given to all European armies from the British army to Russian troops, but there are still many recorded incidents that tell of butts being used in close-quarter fighting. During the American Civil War bayonet practice was taught to recruits straight from the manual and during these periods of instructions on Union officer believed the troops looked like: 'a line of beings made up about equally of the frog, the sand-hill crane, the sentinel crab and the grasshopper: all of them rapidly jumping, thrusting, swinging, striking, jerking every which way, and all gone stark mad.' Given time it would all make sense as the men mastered the techniques and learned how to handle a heavy musket fitted with a bayonet that measured up to 6 feet in length.

In order to prepare them in the eventuality that they may one day face an enemy lunging at them with a rifle fitted with a bayonet, armies devised a series of drills or exercises in order to protect themselves and ward off being wounded or killed. Training instructors at the military depots taught bayonet fighting skills, sometimes known as 'bayonet fencing', using bayonets which had had their points blunted. Even so, receiving a thrust from a blunted bayonet could be painful and in some cases inflict injury. To overcome this problem training instructors left the bayonet scabbard fitted to cover the blade. It was safer but not a foolproof method and injuries were still incurred which were painful more than fatal. This in turn led to the development of special training devices for use in bayonet fighting such as the item invented by John G. Ernst in America known as the 'Improved Bayonet Guard' in 1862 during the country's Civil War. The device comprised of a rubber ball fitted to the 'chape' or tip end of the scabbard to buffer any contact rather than the hard tip of the scabbard. A special fastening clip was also fitted to the scabbard as a safety feature to prevent it from being pulled off during training and prevent the blade from being exposed. Some armies have tried to introduce a high degree of realism by suspending bags of animal blood and intestines obtained from butchers, so that when the bag is punctured by the bayonet the man is faced with a mess resembling a human form. This is an extreme form of psychological training and fortunately is not widespread.

The British army in the second half of the nineteenth century developed a special device, which was a spring-loaded pole to represent a rifle, and was tipped with a rubber ball. Two men were matched against one another to simulate bayonet fighting using these devices. The men wore quilted padded jackets to protect their chests and mesh face masks of the type used in fencing to protect the face were also worn. The men would lunge, thrust, parry and counter using the techniques in the drill manual. It was better than the straw-filled sack technique because the men had to use their skills to attack an opponent and parry thrusts in defence. Such pairing off to practice bayonet fighting was also used by American troops in preparation to going to fight in Europe in 1917. The German Army also used similar training equipment and protected their hands with padded gloves resembling those worn by boxers. If a technique works it will be copied by other armies and this would appear to have been the case with bayonet practice. By the Second World War the large number of troops that had to be trained quickly meant that such techniques could not always be used and the more traditional straw-filled sack suspended from a wooden frame was used. Despite all the training such drills would only be put into practice on the battlefield as a last resort during the Second World War.

In the First World War many senior officers held the belief that no man was bayonetted unless he had put up his hands first in surrender. The ordinary soldier in the trenches would disagree with this opinion and did not relish the idea that there may come a time when he would have to put his resolve to use the bayonet to the test. Those who did stab a man with a bayonet recall how pushing it into a man's body was like pushing a large knife point first into butter. To some that is how it must have seemed, but to others it would have been like stabbing the point of a knife into a lump of raw meat. Some of these men who had used the bayonet on opponents in battle also recall that sometimes withdrawing the bayonet could be difficult as the blade became gripped by muscle. To prevent this from happening training instructors taught the technique of giving the rifle a twist to loosen the blade. Another action to loosen the blade was to fire a bullet, which would enlarge the hole by destroying muscle tissue, and the blade could slide out. Stuart Cloete was a young officer serving in France with the British army and witnessed a remarkably strange incident when one of his men and a Prussian Guardsman charged at one another with fixed bayonets and ran each other through as they closed. Both men were fatally wounded.

In 1917 recruits under training for the British army were instructed in bayonet drills before being sent to the Western Front in Europe. Even at that late stage in the war, when machine guns had halted many charges before the men had advanced, recruits were being instructed to advance until about 200 yards from the enemy lines and then 'go in with the bayonet'. The problem lay in the fact that some of the instructors had never actually been to the Front and did not know what conditions were like. They stuck to the drill manual of the time which lay down that; 'All ranks must be taught that the aim and object is to close quarters with the enemy as quickly as possible so as to be able to use the bayonet. This must become a second nature.' Training was conducted at regimental depots in Germany, Russia, France, and in Britain such as the Guards Depot at Caterham in Surrey. Other training centres were established in France by the British army such as Bailleul, Amiens and the infamous Étaples, south of Calais, which would gain notoriety among the troops for its harsh training methods and become known as the 'Bullring'. Training instructors at such centres would impart such useful information as bayonetting wounded enemy on the ground because they could still throw a grenade or shoot a pistol. Recruits were instructed as to the 'vulnerable points' such as the area of the groin, breast and throat where all the vital arteries are located. To help develop accuracy a small circular ring was attached to the end of a pole and soldiers were expected to be able to lunge through the ring as the instructor moved it up and down and left to right. The soldier was expected to be able to react to any position instantly. This was all well and good on the training ground, but trainees would discover that things were so very much different in the trenches.

Drill manuals of the day stressed how: 'In an assault the enemy should be killed with the bayonet. Firing should be avoided for in the mix-up a bullet, after passing through an opponent's body, may kill a friend who happens to be in the line of fire' Such manuals were in use with all armies training the millions of recruits who would serve in France, Russia, the Middle East and Africa. The instructions were clear in any language that the so-called 'spirit of the bayonet' be inculcated into all ranks 'so that they go forward with all the aggressive determination and confidence of superiority born of continual practice.' A British army instruction manual for bayonet practice emphasized that a man should: 'Go straight at an opponent with the point threatening his throat and deliver the point wherever an opportunity presents itself.' Such opportunities were presented and over in an instant and men reacted accordingly. All armies developed a series of movements to use the bayonet to maximum effect in the confined spaces of the trench systems. These included holding the rifle at extreme length to deliver the thrust and maximum length; short stabbing actions holding the rifle with both hands on the forestock; overhead, upward stabbing movements to deal with men as they leapt across the parapet of the trench, were all instinct actions done in the moment. Australian-born historian and author, the late John Laffin, interviewed hundreds of veterans from the First World War and asked them the question: 'Did you actually wound an enemy soldier with your bayonet?' Many told him that they had not been in a position to do so, and that the Germans frequently ran away or surrendered before the chance presented itself to use the bayonet. Some of his interviewees answered him with a wry smile or a shrug shoulders and a few did admit to having 'stuck the bayonet in.' John Laffin in his writing stresses that such interviews were purely anecdotal and not hard evidence corroborated with further substantial evidence. It was a case of having to take the words of the men. It is unlikely such men would fabricate events.

John Laffin served in 2nd Australian Imperial Force during the Second World War and fought in New Guinea and was wounded. During his basic training in 1940 he was given bayonet instruction by a senior NCO who was a veteran of the First World War who told him to: 'Take no chances, bayonet the bastards as you pass'. His instructor informed him that it was easier to bring oneself to bayonet a fallen enemy than to shoot him. Laffin held the belief that this

Recreated British infantryman with No 4 Lee-Enfield rifle with fixed bayonet demonstrating method of training.

notion was true. In 1943 Laffin devised a style of bayonet fighting which was sufficiently influential in style to be filmed as part of a training programme. His techniques emphasized that a man's hands were his most vulnerable part of his body because they held the rifle. Laffin believed that by 'slashing' a man's fingers or 'sticking' the point of the bayonet into his hands he would be disabled. That is true to a certain degree and Laffin himself recognized this fact and covered the possibility that such actions would not incapacitate a man, and in such an eventuality he should be 'dealt with by bayonet or butt.'

Bayonet practice is a very physical drill and cannot be taught in a classroom like modern techniques using computer simulators. Recruits have to be taken to the training area and shown how to use the bayonet and then he is

Recreated Australian troops from North African Campaign armed with No 3 Lee Enfields and fixed bayonets.

expected to do it himself. Bayonet practice is still conducted in modern armies and it has been proven to be an effective measure in a number of conflicts since the end of the Second World War. The traditional bayonet charge was consigned to the past even before the outbreak of the First World War but the bayonet did not disappear from the infantryman's personal equipment. During the Second World War bayonets were used to intimidate and following the early British and Commonwealth successes against the Italians in the North African theatre of operations images appeared in newsreels of the time showing a few dozen British infantrymen carrying rifles with fixed bayonets escorting thousands of Italians into captivity as prisoners of war. Bayonet practice may seem like it belongs to an age of long ago, but it was the result of such training which allowed the Scots Guards to fix bayonets and storm the Argentine positions on Mount Tumbledown during the Falklands War in 1982. Troops on deployment to more recent areas of operations such as Iraq and Afghanistan frequently patrol on mission with fixed bayonets. This practice continues and because of its usefulness in situations the bayonet will continue to remain as part of the infantryman's basic personal equipment. As a possible result of events in Falklands, where amid all the modern missile weaponry bayonets were used with good effect, the US Army in 1982 returned to the serious teaching of bayonet practice once again as a way of expanding the infantryman's fighting capacity at close quarters. Troops who had never before been taught such combat techniques found themselves learning a new series of rifle drills involving the bayonet and this is continued in all armies today.

Chapter 9

Cut and Thrust

Napoleon Bonaparte may have held a healthy respect for the bayonet and those who were wounded by it and he was also known to favour the use of bayonet in battle, but he would have been surprised at the low numbers of casualties attributed to the bayonet had he cared to study the statistics. A study of the wounded admitted to the Les Invalides Hospital in Paris where disabled veteran French soldiers could retire was conducted in 1715 with the aim of compiling a list of causes of casualties. This revealed that 71.4 per cent of wounds were caused by firearms, mainly muskets, and 15.8 per cent of wounds were caused by swords. Artillery caused 10 per cent of casualties and 2.8 per cent of wounds were caused by bayonets. This is a round figure and gives a total of 100 per cent of casualties in the study. A second study of the wounded at Les Invalides was conducted in 1762 produced an ambiguous result and not unsurprisingly it showed a change in the causes of wounding. The most noticeable effect during the intervening 47 years was the change in the number of casualties caused by firearms, including muskets and pistols, which had dropped to 68.8 per cent. The number of sword wounds had also dropped slightly to 14.7 per cent, a reduction of 0.9 per cent. The numbers of wounds caused by artillery had, on the other hand, increased 3.5 per cent to produce an overall figure 13.4 per cent. This increase indicates the reason for the drop in other causes of wounds. Wounding by bayonets had dropped by 0.4 per cent to give a figure of 2.4 per cent, a fact also caused in part by artillery. The total of the second list was compiled the year before the Seven Years' War ended, produces a figure of 99.3 per cent. From these figures it can be safely assumed that the remaining 0.7 per cent of casualties were caused by multiple wounds or were not conclusively identified. The change in causes of wounding also indicates the changing nature of war with artillery evolving to become the dominant weapon on the battlefield. A similar study conducted 100 years later shows that prior to 1850 swords and bayonets account for around 20 per cent of wounds. After 1860 this figure drops to between four and six per cent and the figures for wounding by artillery increases as does the figure for small arms due to the fact the first machine guns were being introduced at around this time.

It has been opined that the reason why figures for bayonet wounds is so low compared to other causes was because the weapon killed more effectively. This meant that fewer soldiers survived the trauma of being stabbed with a bayonet and never made it to the hospital for treatment. Alternatively, one can also interpret the figures in a different manner and reach the conclusion that perhaps a large number of bayonet wounds were not recognized as being so severe and produced injuries which soldiers could treat themselves and even survive without medical aid, providing the wound did not become infected. A musket ball striking a man could drag pieces of his dirty uniform into the wound, which would cause infection. Also the musket ball would produce an effect called 'cavitation' which pushed muscle tissue apart and allow air to enter which in turn could produce gangrene. Wounding by a blade would produce internal bleeding in the torso and if the intestine or gut was punctured it would cause peritonitis as the body slowly poisoned. A penetrating thrust to a leg or arm was not so life-threatening unless an artery was severed or blood poisoning set in. Even so, the application of a tourniquet could be used to stop the flow of blood by pressure applied to a point by tightening a strap. The instrument was invented in 1718 by a French surgeon by the name of Jean Louis Petit. The device was often used to restrict the flow of blood into a limb during amputation to reduce the loss of blood. The same technique could be used when treating a deep penetrating wound to a leg caused by a bayonet.

The actual point of the bayonet is small but its impact on penetrating any part of the human body, which is its target, will always produce results. The soldier thrusting with the bayonet must do so with the weapon in a line to the centre mass of his opponent and with sufficient force to pierce webbing equipment, uniform and leather belts. Heavy woollen uniforms, metal fastenings such as buckles and leather cross-belts can lessen the impact so the attacker has to push with great weight and strength to pierce such items. A puncture wound caused by a bayonet can penetrate into the lungs to produce a sucking wound in the chest, fracture bones in the arms and legs. Loss of blood through such wounds will weaken one's opponent. For example, deep penetrating wounds causing trauma will affect the olfactory, which controls the sensory system, leading to the loss of control of the bowels and bladder. It is never a pleasant sight to watch the life slowly ebbing out of a man fatally stabbed with a bayonet and so proponents were always encouraged to thrust with all their force, twist to release the blade and move on to continue the momentum. Anyone who has killed an enemy in such a manner is unlikely to ever forget the experience. Even the training in modern armies expounds the quick bayonet thrust, but few modern soldiers have ever used the bayonet in such a way.

The bayonet has grooves formed along the length of the blade and these are called 'fullers' which are designed to allow air to ingress into the wound cavity to prevent any muscle spasm causing a vacuum if there is a contraction. This 'sucking' effect has been dismissed but it can happen in the very large muscles such as the thigh and when it does occur it can prevent the blade from being easily withdrawn from the victim's wound. If the bayonet does become stuck due to the vacuum effect or lodged in bone the simplest solution advocated in training is to fire a round into the body of the victim to enlarge the wound, allowing air to enter and the blade can be recovered. The torso represents the largest target area for a bayonet thrust and with the main vital organs located here it was an obvious part of the body to attack. A man holds his rifle with its bayonet attached at an angle which is directed at this part of the body but he can equally stab at the head, arms, neck, groin and legs in close-quarter fighting and disable the opponent. Records of medical studies reveal that the head is the least likely place to be stabbed at with a bayonet but that does not mean it is never targeted. Whilst the number of bayonet wounds are low compared to other causes the number of head wounds caused by bayonets is very few compared to chest or stomach.

At the Battle of Inkerman it is estimated that only six per cent of all wounds were caused by bayonets, which may not sound very high. The French and British losses combined were 4,676 killed and wounded. The Russians lost 10,216 killed and wounded. When these figures are examined and the percentages worked out this gives a total of 280 and 613 troops respectively who were killed or wounded by the use of the bayonet which is actually quite a high figure. A study of battlefield casualties of the eighteenth century reveals that 40 per cent were caused by musket fire, another 40 per cent was caused by artillery and the remaining 20 per cent was caused by a combination of swords and bayonets. As the historian J.F. Puysuger wrote in his work, *L'art de la guerre par principes at par regles*, which appeared in 1748, 100 years after his father had recorded the bayonet: 'firearms and not cold steel now decide battle.' He continued by suggestion that one should visit the hospital to see how few men had been wounded by the bayonet compared to the musket.

Indeed, when one looks at the above figures one has to agree that he is absolutely correct, but the bayonet could not be dismissed because it could still be used to turn a charge in favour of the attackers to decide the outcome of an assault on positions.

Battlefield wounds occur in three main forms. Firstly, there are compress wounds caused by blunt weapons such as maces and includes crushing wounds from axes. Secondly there are incision wounds caused by swords or knives. Finally there are puncture wounds from arrows and swords or daggers used to stab. Early firearms also added to this range of wounding but the effects of these fell into the puncture wound with some crushing if the projectile struck bone. Bladed weapons existed long before the bayonet, which was simply a development of the knife or dagger. In fact, throughout its history the bayonet has retained it dagger-like shape although there have been variations on the design. Because of the handy utilitarian design troops in their camps have always used their bayonets as tools and socket bayonets in particular made obvious candle stick holders by putting the candle into the socket head. The blade was then pushed into the ground where it could pick up debris, which could later be carried into a wound. The wound may not be fatal but left without medical treatment the long-term effect will result in infection such as blood poisoning and tetanus could set in and lead to death much later.

The French surgeon Baron Larrey reported only having ever treated five bayonet wounds during his long and distinguished medical career, which saw him campaign throughout the Napoleonic Wars from Spain into Russia. This low figure led him to conclude that it was the psychological effect of the weapon, which was sufficient to cause men to run away in order to avoid being stabbed by the blade of 18 inches of steel. A musket ball meant a quick death in most cases but a stab wound by a bayonet was a slow lingering agony as one bled to death. A Surgeon-General serving in Napoleon's army believed that for every bayonet wound treated there were a hundred more caused by artillery or musket fire. In other words, only one per cent of all wounds were caused by bayonets. Despite it being carried by all infantry armed with muskets, there were some who agreed with Larrey's opinion that bayonets were very rarely used and of limited usefulness except when storming fortified places or in isolated skirmishes. In fact, Larrey studied the numbers of wounded following a typical engagement during the Russian campaign and found that 119 wounded or 96 per cent of the total figure were wounded by musket fire and only five wounded or 4 per cent of the total studied were wounded by bayonets. Other leading medical men serving with the military such as the British surgeon George J. Guthrie who served with Wellington on campaign in Spain during the Peninsular War asserted that regiments 'charging with the bayonet *never* meet and struggle hand to hand and foot to foot; and this for the best possible reason, that one side turns and runs away as soon as the other comes close enough to do mischief.' The French General Baron Lejeune supported this idea and observed on bayonet charges that they were 'very rare in modern warfare, for as a rule one of the corps is demoralised to begin with by the firing, and draws back before the enemy is near enough to cross muzzles.' This group of people thought that in battle it was the merit which created an entirely psychological effect and that bayonet charges were normally only made when the enemy was already wavering as a result of artillery fire or musketry, which is exactly what Lejeune had expressed.

Guthrie admits to treating very few cases of bayonet wounds in his career. Indeed, medical notes on the treatment of such wounds is rare but it is possible to draw comparisons concerning the effects between bayonet wounds and those inflicted by other bladed weapons such as swords or lances. One case recorded by Guthrie concerned an unidentified soldier who had been 'wounded in the right side of the chest by a sword, which has passed slantingly under the shoulder' and was typical of the wound inflicted on an infantryman by a cavalryman. The patient's chest became swollen to the point where he was struggling for breath. Guthrie opened the man's wound again and evacuated the build-up of air, and he 'could distinctly hear the air rush out.' The man recovered. Another patient he treated had been caught in the left side between the fifth and sixth ribs. He too suffered from swelling in the abdomen, which was made more comfortable when Guthrie opened his chest to release the air. Again, the man went on to a full recovery. Guthrie observed that bayonet wounds to muscular parts, such as arms and legs, often healed with little trouble. Should such a wound require treatment he advised that it be cleaned with water and bandaged but not with a compress dressing. Having seen many types of puncture wounds he concluded that in his opinion lance wounds were not as dangerous as those inflicted by bayonets. Presumably he reasoned that due to the fact a bayonet had been used for many purposes it would be dirty and therefore likely infect a wound more readily.

Surgeons on being presented with deep penetrating wounds caused by bayonets would have realized there was little, if anything, they could do.

Superficial wounds to the abdomen caused by bayonets would have been bandaged or sutured if required. Such wounds to the arms and legs may have been cauterized to stop bleeding and then bandaged and were considered to have been non-threatening to life unless dirt entered the wound to cause infection. Surgeons such as Guthrie or the French Jean-Baptiste d'Hèralde would have been experienced enough to recognize internal bleeding from a major organ caused by a bayonet thrust and would have known that apart from making the patient comfortable there was little they could do. A deep penetration would leave few survivors. Surgery to suture arteries was possible at this time but deep invasive remedial surgery was years away and men wounded in such a way would have passed out through loss of blood long before they bled to death.

The bayonet is designed to inflict puncture wounds but these are not necessarily instantly fatal unless inflicted on the heart. For the most part bayonet wounds are long, lingering deaths as the wounded soldier bleeds to death slowly. If an artery was punctured in in battle by any weapon prior to the nineteenth century it invariably meant death unless medical treatment was close at hand and if the surgeon was competent. By the twentieth century medical procedure had advanced so that treatment to puncture wounds, even to an artery, could stem the bleeding. Vital organs deep inside the abdomen were vulnerable to all weapons on the battlefield especially bayonets and wounds to the stomach, kidneys or liver would invariably prove fatal. Before the advent of antibiotics, such as penicillin, there was little that doctors could do, beyond using vinegar as an antiseptic, to prevent a wound from becoming infected which would lead to blood poisoning or tetanus inflicted by bayonets with dirty blades. Pieces of dirty uniform dragged into wounds caused by bullets were also a problem which caused infection, but it was the bayonet which had been used for all manner of purposes such as digging, and the fact they would not have been cleaned properly after an earlier battle, which was the single most common source for producing battlefield wounds infected with bacteria.

During the Peninsular War the British made a night attack against the walled city of Ciudad Rodrigo on 19 January 1812. As they advanced the troops were reminded by their officers: 'Recollect you are not loaded! Push with the bayonet.' General Picton in particular addressed the 88th Regiment of Foot by saying: 'It is not my intention to expend any powder this evening. We'll do the business with cold iron.' Among those who charged the position was Lieutenant

William Grattan, who later described the actions of Sergeant Pat Brazil and Privates Swan and Kelly of the 88th Regiment of Foot (later to become the Connaught Rangers which would be disbanded in 1922). Lieutenant Grattan wrote how he saw the men jump into a gun-pit defended by French troops: '[they] engaged the French cannoniers hand to hand, a terrific but short combat was the consequence. Swan was the first, and was met by two gunners on the right of the gun but, no way daunted, he engaged them, and plunged his bayonet in the breast of one; he was about to repeat the blow upon the other, but before he could disentangled his weapon from his bleeding adversary, the second Frenchman closed up on him and by a coup de sabre severed his left arm from his body, a little above the elbow; he fell from the shock, and was on the eve of being massacred, when Kelly, after having scrambled under the gun, rushed onward to succour his comrade. He bayoneted two Frenchmen on the spot, and at this instant Brazil came up; three of the five gunners lay lifeless, while Swan, resting against an ammunition chest, was bleeding to death… Brazil… in making a lunge at the man next to him… slipped on the bloody platform, and he fell forward against his antagonist, but as they both rolled under the gun, Brazil felt the socket of his bayonet against the buttons of the Frenchman's coat.' This excerpt tells how desperate close-quarter fighting was with no mercy being shown. It was a case of kill or be killed with any hesitation being seen as weakness and almost certainly leading to death. There was no time to think in such circumstances only action and reaction.

Whilst bayonets were used in close-quarter engagements they were primarily used to protect the infantry from attack by cavalry. In this role the infantrymen lunged at either horse or rider. A man would thrust at whichever presented itself as the most vulnerable target. The rider was usually engaged at very close quarters and the horse being the largest target was always going to be attacked. For example, William Tomkinson was a rider during an attack by a British cavalry unit against positions held by French infantry in 1809. Tomkinson was shot through both arms just as he was about to fire his pistol at a man. His horse was then bayoneted, which as Tomkinson later stated prompted the animal to charge off: 'full gallop to the rear and coming to the fence of an enclosure he selected a low place in it under a vine tree, knocked my head into it, when I fell off him.' The animal was obviously panic-stricken from the noise and pain and being wounded in both arms meant he could not control the animal, which charged away out of control. The rider was more fortunate than many who were killed in similar conditions.

Despite the observations of experienced officers such as the French General Louis Trochu who in the nineteenth century proclaimed that in his entire military career he had only ever seen three engagements where bayonets were used. Of these he believed that the incident he witnessed at the Battle of Inkerman, 5 November 1854, during the Crimean War, when the French 3rd Chasseurs-à-Pied and 6th Infantry of the Line crossed steel with a force of Russians, was a chance encounter when troops collided in the fog. Lieutenant Henry Clifford was serving as aide de camp to Brigadier General George Buller with the 77th Regiment of Foot (later to become the 2nd Battalion of The Middlesex Regiment) and together they were making their way through the fog towards the sound of the firing. Suddenly, through a gap in the fog, Clifford spotted a group of Russians moving forward and ordered: 'In God's name, fix bayonets and charge.' A desperate fight then ensued during which Clifford drew his revolver. He later recalled how: 'the brave fellows dashed in amongst the astonished Russians, bayoneting them in every direction. One of the bullets in my revolver had partly come out and prevented it from revolving and I could not get it off. The Russians fired their pieces a few yards of my head, but none touched me. I drew my sword and cut off one man's arm who was in the act of bayoneting me and a second seeing it, turned round and was in the act of running out of my way when I hit him over the back of the neck and laid him dead at my feet.' This was another chance encounter in the same battle rather than a true bayonet charge as stated by General Trochu, but there is no doubting in either case that the bayonet proved most effective in such close-quarter fighting. Even so, it would appear that encounters were not decided by the use of the bayonet and the study of wounded, after some engagements, proved how low the figures were for bayonet wounds. For example, after a typical engagement a study of 720 wounded revealed that seven were caused by sword and lance, with one man later dying. There were 36 wounded by being bayoneted in the same engagement, which equated to 0.5 per cent of the total figure and of these four later died.

There are accounts which throw a different light on bayonet fighting and its effectiveness, especially at close quarters. In 1813, for example, during the Peninsular War, Corporal Wheeler of the 51st Regiment of Foot happened upon the aftermath of an engagement between Spanish and French infantry. He recorded that: 'a desperate job it must have been and no mistake about it, for the contending parties lay dead bayonet to bayonet. I saw several pairs with the bayonet in each other.' Benjamin Harris who served with the

95th Regiment had never seen bayonets crossed, when he came across scene where a private of the 43rd Regiment of Foot and a French infantryman had bayoneted one another simultaneously and could not help but study it 'with much curiosity'. Not all bayonet wounds were fatal and if they did not become infected there was no reason why a man should not recover and continue to fight. For example, John Cheshire from Stockwell served with the British Legion, which was raised to fight in the First Carlist War in Spain between 1832 and 1839. He survived the war and he was eligible for a pension and on his certificate it is listed that he had received '35 bayonet wounds in various parts of his body'. He was extremely lucky because none of the wounds became infected or had pierced a vital organ or artery.

During bayonet practice recruits are told how at times it can be difficult to extract the bayonet from after stabbing into the body. This is usually caused by the blade sticking into bone, but sometimes the flesh can cause a 'sucking' action as it closes around the blade. In such circumstances the recruit is instructed to fire a shot to enlarge the wound to release the blade. This may sound cold-blooded and a story made up by training instructors, but the bayonet can become stuck on occasion. For example, Private Maxwell of the 51st Regiment of Foot (later to become the King's Own (Yorkshire

Battle of St Jean de Luz, 10 December 1813.

Light Infantry)) related how he bayoneted a French dragoon with such force that his bayonet became stuck, and he could only extricate it from the man's body by standing on his chest and pulling vigorously.

A case study was made of 144,000 troops who had been killed during the American Civil War for whom the cause of death was known and it revealed that 108,000 had been killed by rifle fire. A total of 12,500 had been killed by artillery and 7,000 killed by sword or bayonet. The remaining 16,500 had presumably been killed as the result of other factors or a combination of wounds caused by rifle fire and bayonets. During the American Civil War 622,000 soldiers were killed on both sides, so this case study only represents just over 20 per cent of deaths between 1861 and 1865. Losses in the First World War were horrendous with the artillery accounting for the greatest proportion. Small arms fire, hand grenades and mortars accounted for more and gas killed thousands, also. An assessment of the overall casualty figures reaches the conclusion that bayonets only accounted for an estimated 0.3 per cent of all battlefield wounds. Bayonet charges in the face of machine guns firing from well-sited positions were useless and yet they were still ordered. These head-on assaults were not so much bayonet charges but attacks by infantry in an attempt to neutralize the machine guns and were carried out by men armed with rifles with fixed bayonets.

Chapter 10

Ceremonies and Other Weapons Fitted with Bayonets

Rifle drill with bayonets fixed.

As long as there have been armies there have been ceremonial parades to commemorate a victory or honour a leading figure. The Legions of Scipio, Julius Caesar and Tiberius paraded through the streets of Rome to mark their victories on their return from campaign. Over 2,000 years later the armies of Hitler were marching through their newly won conquests in similar victory parades. Armies marching off to war display their smartness and with rifles carried on their shoulders they usually have bayonets fixed. Outside palaces and castles soldiers can often be seen on parade with fixed bayonets

US Marines on deck of warship with bayonets fitted.

French troops 'present arms' to an unidentified general officer. The rifles are Lebel and the bayonets are the 'Rosalie' type used during the First World War.

Street lining for Coronation of King George V. The guard of honour have bayonets fixed.

Print showing British
troops in 19th century
with bayonets fixed to
weapons.

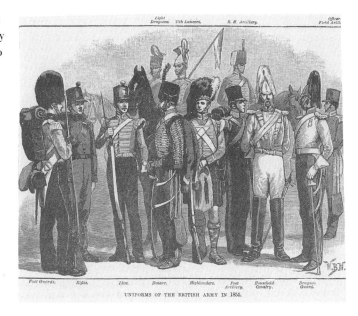

Tsar Nicholas II of Russia leaving Compiègne, 1901. French troops have bayonets fixed for
salute.

to demonstrate drills. For example, outside the Kremlin in Moscow, Russia, the guards exercise immaculate drill with rifles and bayonets attached. In England similar scenes are to be found at locations such as Windsor Castle in Berkshire, Buckingham Palace and St James's Palace in London along with the Tower of London. The order to 'Fix bayonets' is more likely to be heard on the drill square rather than on the modern battlefield. When it is given the troops move in unison to pull their bayonets from their scabbards and then move to fix them to their rifles. The movement transforms the unit from looking just smart to looking impressive.

The ceremony of Changing of the Guard involves rifle drill with bayonets attached and the Trooping of the Colour each year in London also involves such drills. Armies around the world perform ceremonies known as 'street lining' for the benefit of visiting dignitaries and involves Guards of Honour parading along the route taken. As the cortege passes each section the soldiers present arms in salute as a mark of respect and the bayonet is always evident. Sentries outside Government buildings during time of war also paraded with fixed bayonets but this role was more serious and intended for security measures. Ceremonial parades involve troops using drill movements to give

French Colonial troops charge using bayonets at the Battle of Solferino in 1859.

demonstrations of skills using rifles with fixed bayonets. Some countries select the very smartest men to form in special rifle drill display teams such as the 3rd US Infantry Regiment (The Old Guard). This is the official Escort to the President of America and has a history stretching back over 200 years being engaged in many campaigns and today it still serves in war zones such as Iraq. As well as being fighting soldiers, members of the Drill Team participate in military tattoos such as the world-famous annual Edinburgh Tattoo in Scotland. The members of the team perform a series of complicated drill movements using rifles with fixed bayonet in complete silence.

In the past all infantry troops were issued with bayonets which they carried on waist-belts, and even attached units serving in specialist roles such as engineers and sappers were issued with bayonets which had been specially designed for use in heavy-duty work. Although these were primarily intended for being fixed to the end of a rifle these bayonets were more often used as tools. Some bayonets were more for ceremonial show as noted in a Royal Warrant from 1781 which decreed that 'the drums and fifes' in the British army should carry short-swords which should have 'scimitar blades' and reflects certain other eastern influences which the army had adopted such as turban headdress. Today drummers and bandsmen are musicians but they still carry short-swords or bayonets, referred to as 'side arms', which are worn on their parade belts. These short swords or bayonets are symbolic and form part of the overall uniform code of dress for drummers in many armies around the world today and are worn on ceremonial parades such as Trooping the Colour in London. In the British army line regiment bandsmen were at one point issued with the 'Sword, drummers, Mk II' which was usually carried in a black

Recreated British troops of the nineteenth century showing method of stacking muskets in camp with fitted bayonets.

German FG42 rifle with 'spike' bayonet as used by the Fallschirmjäger (parachute troops).

leather scabbard suspended from a white buckskin frog attached to the waist belt. Some examples were highly decorated with lion-head pommels or the Royal Cypher of the monarch. Towards the end of the nineteenth century, around the time of the second Boer War, 1899 to 1902, regimental bands of the British Army had stopped parading in the actual battle area and the drummers and musicians were used to serve in the role of stretcher-bearers. This additional duty was observed in other armies and had long been a traditional role for troops who were otherwise non-combatant. Even as stretcher-bearers the drummers continued to wear their short-swords and in 1901 a List of Changes issued for the British army stated that all drummers' swords were to be sharpened before going on service. This order may have stemmed from an alleged incident concerning the massacre and mutilations of boy drummers, which was reported to have occurred during the Zulu War of 1879. Troops know that although the bayonet is used as part of ceremonial duties it is first and foremost a weapon of war.

The bayonet as a weapon is intended to be used with the standard issue service rifle in an army, but over the years it has been subject to change, which includes being modified for use with other non-standard weapons. Armies are also given to creating specialist units to serve in a specific role for which the troops are trained to a very high standard. These are known as 'Special Forces' and during the Second World War included units termed as Commandos and airborne troops. Flying troops into a combat area and delivering them by parachute had been thought of during the First World War but there was no aircraft capable of operating in such a role. By 1939 countries such as Russia had demonstrated how parachute troops could be dropped in large numbers to quickly seize an area and hold it until heavier units came to its support. The idea was not wasted on Germany where General Kurt Student built up the Fallschirmjäger. They carried standard infantry weapons such as the K98 rifle and bayonet, but later some specialist weapons were later developed. One of these was the Fallschirmjägergewehr 42 or FG42. It was produced in two main designs, which resembled a light machine gun, and with a cyclic rate of fire between 500 rounds per minute and 750 rounds per minute it was equipped with a bipod. It was also equipped with a bayonet that resembled a rod and when not in use it was carried reversed under the barrel. The spike-like blade was cruciform in design, like the Russian Mosin-Nagant rifle with the Type 1 bayonet measuring 11.8 inches in length and the Type II slightly shorter at 10.4 inches in length. Apparently a later version only 6 inches in length was also produced but the idea was never any good. The FG42 measured between 37.2 inches and 38.4 inches overall and between the two main versions only some 15,000 weapons were produced. Parachutists had little need for a bayonet due to the nature of their role but they carried

Recreated Soviet troops with man on left armed with a Mosin-Nagant rifle with folding bayonet.

them all the same, but on the FG42 the bayonet was never going to amount to much. The Japanese developed the Type 100, which used a similar style spike bayonet stored under the barrel, but records show that very few of these were actually produced.

The Italian Army fitted the MAB 38A sub-machine gun with a bayonet and Australia developed two designs of sub-machine guns during the Second World, the first of which was the Austen and appeared from around 1941. It fired a 9mm calibre round and measured 33.25 inches with the butt stock extended. It was developed for use in jungle fighting and weighed 8lb 12oz and fired at the cyclic rate of 500 rounds per minute. A small bayonet was developed for it but with a production run of only 20,000 the combination never made any impact. The other Australian SMG design was the Owen, which also fired a 9mm calibre round at the improved cyclic rate of 700 rounds per minute. The weapon weighed 9lb 5oz with an overall length of 32 inches. This was meant for jungle fighting and was fitted with a bayonet weighing 12oz and 10 inches in length. Around 50,000 of these weapons were produced by 1945 and remained in service until the early 1960s. Again such a combination had a limited value because no soldier mounts a bayonet charge using sub-machine guns.

The British firearms manufacturer of Sterling based in Dagenham in Essex produced an SMG called the Lanchester from around 1941 for use by the Royal Navy. It fired a 9mm calibre round at the cyclic rate of 600 RPM and weighed 9lb 9oz with an overall length of 33.5 inches. Fewer than 100,000 of these weapons were produced and numbers of the Pattern 1907 bayonet as used on the No 1 SMLE rifle were modified to be fitted to the weapons. This bayonet had an overall length of 21.75 inches with a blade length of 17 inches, which meant a potential reach length of 50 inches if used as a bayonet combination. It was used by shore parties but it was never a popular weapon because of its weight and bulky size. The US Army and Marine Corps used shotguns fitted with bayonets during the 'Island Hopping' campaign against the Japanese where each position had to be cleared out literally by hand. The Japanese believed in the use of the bayonet to the degree that the Type 96 machine gun was capable of being fitted with the standard infantry bayonet as used on the Arisaka rifle. It fired 6.5mm rounds at the cyclic rate of 550 RPM and weighed 20 lbs unloaded and measured 41.5 inches. The standard infantry bayonet with its 15.75-inch blade as fitted to the service rifle was capable of being fitted to the Type 96 machine gun. This would have given

Japanese Type 96 machine gun with bayonet.

it a potential reach of 57 inches when thrusting but its great weight and bulk would have made it almost impossible to use except from the waist during a massed 'Banzai' charge. The combination was not really a requirement, but rather as a feature that was there should it be needed, as opposed to needing it and not having it. Even so, such a combination was un-necessary and no other light machine gun of the WW II or indeed any other period appears to have been fitted with such an application for a bayonet.

Bayonets continued to be used on shotguns by the Americans during their campaigns against the Japanese during the Second World War and even in post-war years some weapons designers still insisted on including a fitting for a bayonet to certain types of weapons even though these styles of weapon clearly did not warrant such a feature. The category of weapon which this applies to is the classification known as sub-machine guns which includes many well-known types such as the Israeli 'Uzi'. This weapon first appeared in 1950 and since then an estimated ten million have been produced in several versions and used by a number of armies including the Israeli armed forces. A bayonet was developed for the Uzi which had a spear point but other types have also been used which have drop point tips however the compact overall length of the UZI even when fitted with a bayonet renders it an impractical

Yugoslavian Type 56 SMG with bayonet.

combination except for intimidation reasons when guarding prisoners. The Rexim sub-machine gun was a 9mm calibre weapon produced in Switzerland between 1955 and 1957 during which time several thousand were produced. It had a folding skeleton butt stock and measured 24.35 inches with the butt stock folded. Originally it had a spike-type bayonet that was carried stored reversed under the barrel when not in use in the same manner as the wartime German FG42 used by parachute troops. Production was continued in Turkey

Irish Guards in Aden in 1966 with Sterling sub-machine gun, which could be fitted with a bayonet.

where the weapon was modified and fitted with a fixed wooden butt stock and the bayonet was changed to the removable knife-type with a drop point. This bayonet was carried in a scabbard on the user's waist belt, but again such a compact combination had a limited range of use in a combat situation either as a defensive weapon or to guard prisoners.

The Yugoslavian weapon manufacturing company of Crvena Zastava began producing a sub-machine gun in 1956, which was termed the M56 and fired a 7.62mmX25mm calibre round at the cyclic rate of 600 RPM. It was an effective and reliable weapon measuring 23.3 inches with the butt stock folded and 34.4 inches with it extended. It weighed 6.61lbs and was a copy of the wartime German MP40. A drop point style bayonet was developed for it but like all SMGs the role of such a combination was limited. The Portuguese based the development of their FBP SMG on the German MP40 and the American M3 'Greasegun' and fitted it with a bayonet but it was never a popular design. The Sterling SMG was developed in 1944 and manufactured at the Dagenham-based plant in North London after which the weapon was called. During its service life, which lasted into the 1990s, the Sterling SMG hardly changed its appearance and even the compact versions were scaled-down models of the original. Fitted with a folding butt stock like most SMGs it fired a 9mm calibre round at the cyclic rate of 550 RPM. It was used in many campaigns fought by the British Army including Malaya, Cyprus, Northern Ireland and the Falkland Islands in 1982.It was sold to many overseas countries and could be fitted with a drop point knife-type bayonet which was carried in a scabbard on the user's waist belt when not in use. There were few opportunities to use such a combination and although the Sterling is sometimes seen in photographs with the bayonet fitted it was almost certainly never used in combat. Indeed, in the aftermath of the Falkland Islands War British troops of 2 Parachute Regiment armed with Sterling SMGs guarded hundreds of Argentine prisoners of war but they did not fit bayonets to their weapons. No doubt weapon designers will continue to develop SMGs which have the facility to accept bayonets, probably more for the benefit of the users rather than being of any real practical value.

Over the centuries each country has adopted its own specific range of weaponry for its army and this policy is the same from artillery down to the basic infantryman's rifle. Some are fortunate enough to be issued with a type of weapon that has been developed by designers in their own country while smaller countries purchase what they can afford from the larger states

Scots Guards in Kenya with Sterling sub-machine guns with bayonets fitted.

and become their client states. In the case of countries which have enjoyed an overseas empire such as France and Britain, they influenced the armed forces in the defence of these territories. Some weapons have been fitted with integral bayonets a design which had first appeared around the eighteenth century. By the nineteenth century, apart from a few still being produced for a limited demand, the design had all but run its course and the idea fell out of fashion. Short-barrelled weapons known as blunderbusses were popular with armed guards riding as escort on mail coaches to protect them against highway robbers. Some of these weapons were fitted with spring-loaded bayonets and some designs of pistols also incorporated folding bayonets as short, stabbing weapons for self-defence. In 1803 Mr J.S. Searles was granted Patent Number 2744 by the Patent Office in London for his idea for a: '[bayonet] to slide up and down the barrel by means of a piece of iron fixed on the barrel and is catched at each point by a spring and lever'. Beyond this not much is known about the device. Another idea was proposed in 1808 by Mr G. Richards who was granted the Patent Number 3155 for his design which allowed the bayonet to 'slide outside a piece of the barrel when unfixed, with a spring to allow it to slip over the sight at the top of the piece.' The patent describes it 'which spring is to prevent their being unfixed improperly.' When not required the bayonet is folded back over the fore-stock. Unfortunately, no drawing exists of this idea and it would be safe to assume it was never adopted by the military.

Such designs were of little or no interest to the military in the first half of the nineteenth century, but that did not stop designers from registering their ideas with the Patent Office in London. For example, in 1849 L.A. de Chatauvillard was granted Patent Number 12613 for his design for a bayonet that is: 'hinged to the barrel and is turned forward when fixed and retained by a spring.' In 1852 Mr J.J.H. MacCarthy was granted Patent Number 519 for the: 2... bayonet with cylindrical socket slides up the barrel and is locked with a spring catch'. Two years later in 1854 Mr. C. Crickmay was granted Patent Number 2473 for his design for a bayonet which was hinged underneath the barrel and held by a catch which was actuated by a 'trigger'. When the trigger was activated the bayonet flicked forward by means of a strong helical spring and secured in place by a spring catch. Applying pressure released the blade and it was folded back. The Dutch army produced a variation of the integral bayonet by developing a type which had a swivel action and was incorporated into the M1826/30 musketoon. It had a triangular blade in section and the musketoon was a shorter weapon than the standard cavalry carbine and was often issued to Dragoons for use at short range.

Integral bayonets were investigated again by various armies towards the end of the nineteenth century. The American army, for example, developed one type for use with the Springfield M1884 rifle. This could be extended out to a full length of almost three feet. It resembled a ramrod, tipped with a point, and was carried in place of the ramrod. A small release catch fitted under the barrel allowed the rod bayonet to be slide forward telescopically to the length required. It was not a successful design and would have been prone to bending if used with great force. It was tried again on the Springfield M1888 but the rifle was not widely adopted for military use, and the bayonet style was dropped never to be repeated. The Italian army developed a folding bayonet for its Mannlicher-Carcano M1891 Carbine, which would be fitted with various types of blades and release catches throughout its service history, which extended into the Second World War. The later Moschetto Modello 91, 'per truppe speciali' (Mod 91 TS) was equipped with a detachable bayonet fitted with a transverse fixing slot across the back of the pommel. This slotted into a transverse lug mounted

Scots Guards in Nairobi, Kenya, armed with L1A1 rifle with bayonet fitted.

Scots Guards leaving for Borneo. They are carrying L1A1 rifle with bayonets in scabbards.

Irish Guards in Malaya 1964. They carry the Sterling SMG and L1A1 rifle, both of which could be fitted with bayonets.

below the weapon. It is believed this method was developed in the unlikely event that an opponent would try to snatch the bayonet from the rifle.

The Japanese army also adopted a folding-style bayonet in 1911, which was fitted to the Arisaka Type 44 carbine. The Japanese also used the traditional long-bladed knife-type detachable bayonet on other service rifles during both world wars. The Type 44 rifle with its folding bayonet was used during the First World War but had been replaced by the time of the Second World War and the Soviet Red Army used a similar design which was fitted to the 1944 carbine and folded back along the barrel. This had a blade that was cruciform in section and produced a fierce puncture wound. Russian bayonets had this style of blade, which was distinctive and identified them as being used by the Soviet Red Army. Indeed, when the site of a massacre of perhaps as many as 22,000 Polish army officers, police and intelligentsia was discovered at Katyn by German troops in 1943 forensic examinations revealed gun shots and wounds inflected by bayonets with cruciform blades. To many this proved conclusively that it was Russian troops who had committed the atrocity.

In post war years the folding bayonet continued in limited service. The Soviets developed the AK-47 and SKS automatic rifles both of which could be fitted with the standard knife-type bayonet. The Chinese copied these weapons, referring to them as the Type 56/1 and Type 53 respectively and fitted them with folding bayonets, which had blade lengths of 8.7 inches. The Czechoslovakian army as it was then took into service the 7.62mm calibre Samonabiject Puska VZ52/57 automatic rifle and this was also equipped with a folding bayonet fitted to the side of the fore stock. This could be folded out like all other types of integral bayonets and simply clipped back into place when not required. This type of weapon was eventually replaced by the AK–47 equipped with the detachable bayonet of standard knife design. In the 1970s the then country of Yugoslavia adopted the M70 version of the AK–47 for service use and this too had an integral bayonet which folded back on the underside of the barrel and fore-stock. However, whilst it may seem that a bayonet already fitted for use is a good idea it does have its drawbacks. The main problem to an integral bayonet is that if it becomes damaged it ceases to function properly and can no longer be deployed if the spring catch is broken or folded away if the blade becomes bent. The integral bayonet also meant more time producing the weapon and rifle. In the end the design did not become universally accepted and most armies prefer to use the detachable bayonet.

Chapter 11

The Twentieth Century and Beyond

In Europe at the start of the twentieth century there were still to be found adherents to the 'spirit of the bayonet' such as a certain Colonel Black serving in the German Army who in 1911 expressed his belief that: 'a soldier should not be taught to shrink from the bayonet attack, but to seek it. If the infantry is deprived of the arme blanche, if the impossibility of bayonet fighting is preached, and the soldier is never given an opportunity in time of peace of defending himself, man to man, with his weapon in bayonet fencing, an infantry will be developed, which is unsuitable for attack and which, moreover, lacks a most essential quality [namely] the moral power to reach the enemy's position'. Here was a man who still believed in the bayonet charge even though recent experiences in wars since the Franco-Prussian War of only forty years earlier proved that modern weapons could devastate an army before it had advanced very far across the battlefield.

Artillery, machine guns, powerful rifles now dominated the battlefield and even a simple thing like ordinary wire could prevent infantry from achieving its mission as had been demonstrated during the American Civil War. Barbed wire which comprised of strands of wire into which was interwoven sharp barbs to create an obstacle to prevent free movement would prove a much worse obstacle. The first patent for the material was granted to Lucien

Studio photograph of British lieutenant produced as postcard to send home.

Smith of Ohio in America in 1867 but similar material had been used earlier and although designed to keep animals penned, its usefulness as an obstacle was soon realized by the military. American troops used it during the Spanish-America War and it was also used during the Russo-Japanese War. During the First World War millions of miles of barbed wire would be produced and armies would form it into impenetrable barriers many feet thick and as tall as a man to protect their trenches from assault by enemy infantry.

When the First World War broke out in August 1914 all European armies engaged in a period of the initial manoeuvrings in the opening months of the war. Finally things ground to a halt and the opposing sides began to dig in along lines of trenches, which would eventually stretch almost 500 miles from the Swiss border to the Channel Coast. Each army used mile after mile of barbed wire strung onto wooden posts to create a defensive barrier to protect against attacks. The obstacles were laid out in such a way that pathways were left through the wire so that the defenders could move forward to attack the enemy positions. These lanes were tempting avenues for the enemy to advance down and these became known as killing zones because they were commanded by machine guns which would devastate any attack coming along such lines. Artillery was often used in an attempt to destroy barbed wire. As a result, new tactics evolved known as a rolling barrage whereby the artillery would fire their shells and gradually increase their distance. The resulting bombardment could then be used as a 'shield' for waiting troops to advance behind. This was meant to be the opportunity for infantry to advance with bayonets fixed, but despite the reassurances that the enemy would be destroyed or taking shelter from the barrage, such attacks always met with resistance.

On the 1 July 1916 in an effort to break the stalemate of trench warfare the British mounted an offensive in the Somme region of France. Its objective

Bayonet drills were very similar in approach. The British 'on guard'.

was to seize German positions and advance forward. At the end of the first day the British Army suffered almost 60,000 killed and wounded, even though artillery was meant to have neutralized German positions. At the end of a week of fighting the British artillery had fired some 1.75 million shells of all calibres, but still the Germans held their positions. There was opportunity to use the bayonet during the battle as recalled by Lance Corporal Heardman of the 2nd Manchester 'Pals' Battalion who remembered how he came: 'face to face with a great big German who had come up unexpectedly out of a shell hole. He had his rifle and bayonet "at the ready". So had I,… Then almost before I had time to realise what

Downward thrust.

was happening the German threw down his rifle, put up his arms and shouted "Kamerade"'. Things happened quickly indeed and this incident, whilst isolated, does show the attitude of the infantry towards the bayonet when in close contact and contradicts Colonel Black's emphasis on the use of the bayonet to achieve the objective and drive the enemy away. Machine guns and artillery were certainly proven as dominant weapons, but on the battlefields of the First World War they were to be joined by other weapons such as aircraft and Zeppelins for long-range strategic bombing along with armoured vehicles including tanks. Science also added to this array of weaponry and flamethrowers were used to burn up whole stretches of trench systems. Then, in early 1915 the Germans added another dimension to all this by releasing poison gas which was indiscriminate.

To the men in the trenches they believed it could not get much worse, but the war was about to enter a new phase. Between April and May 1915 the Germans released tons of poison gas, which killed hundreds and incapacitated thousands more by its effects. This was the first use of a weapon

Extended thrust gave longer reach. However, it was unstable and recovering the weapon took more time.

Full weight was put behind such a movement.

of massed destruction and the troops gave it a name 'Frightfulness' because it was terrifying. Every main army in WW I used poison gas but despite its fearsomeness its use has been estimated to have killed only some 91,000 men and incapacitated to varying degrees over 1.2 million men in all armies. Whilst this figure is a great deal higher than the casualties caused by bayonet wounds it is much less than the cause of death or wounding due to either artillery or machine gun fire. During the course of the war there would emerge a coalition of twelve main Allied nations which between them mobilized over 42,188,000 troops. Of this figure almost 4.9 million were killed and a further 12,809,280 were wounded. The countries which formed the Central Powers, Germany, Austria-Hungary, Turkey and Bulgaria, collectively mobilized 22,850,000 troops, of which almost 3,132,000 were killed and a further 8,419,533 were wounded. Some sources claim that a figure as low as perhaps only 0.3 per cent of all wounds were caused by bayonets during the First World War. Indeed, the writer and historian Thomas H. Wintringham in his book *Weapons and Tactics from Troy to Stalingrad*, states that 1.02 per cent of casualties in the war were caused by what he describes as 'miscellaneous' wounding of which perhaps 0.32 per cent were caused by bayonets. Presumably the other causes were barbed wire and weapons used during trench raids. If the figure of 0.3 per cent is taken as a conservative total for the number of those killed and wounded by bayonets one arrives at 87,600 for the whole of the war. This represents 9,300 killed and 25,200 wounded from the Allied casualty list. For the Central Powers the figures for those killed and wounded is 14,700 and 38,400 respectively. Given the reputation of the bayonet and its use as a personal weapon these figures are remarkably low and really confirm the observations of surgeons such as Larrey and Sir Charles Bell from an earlier period.

There was no way the troops serving in the trenches would have known that the bayonet was causing so few casualties. To them they were 'getting to grips' with the enemy and even had they known the bayonet was only causing

3rd Regiment of Line, 1st Bn, 2nd Company, Belgian Army. Their bayonets have relatively short blades.

Belgian Caribineer with bayonet fixed.

Belgian infantry in Blue Full Gear with bayonet fixed.

a superficial number of casualties they knew artillery and machine guns were killing far more. In May 1915 Stanley Casson was a young officer serving in the East Lancashire Regiment at the Frezenberg Ridge in Belgium and later recalled his experiences in the trenches in his book 'Steady Drummer'. In this work he remembers how the men: 'moved cautiously… still submerged in inky darkness… the order for silence was given, if indeed one can silence the combined clashing and clanking of several hundred bayonets against entrenching tools, of rifles against packs, and of bodies jostling and stumbling against each other.' Having set the scene of events he continues: 'Then came the order to fix bayonets, which was done with a sound like a thousand tin cans being bashed merrily.' It would have been an ominous order to receive but in his account Casson relates they were not alone because there were other battalions of famous regiments in the trenches with him and his men. His reference to the noise of fixing bayonets would have been a sound which would have been heard in trenches held by French, German and Belgian troops and later Americans when they entered the war. Yet, amongst all of the advances in weaponry the bayonet remained as the infantrymen's most reliable weapon next to his rifle. Factories in all the belligerent nations produced millions of bayonets to equip the soldiers being sent to the battlefields. The British company of Wilkinson–Sword produced more the 2.5 million bayonets alone during the war. If one uses the ratio of 5 per cent were saw back types, then some 125,000 from this production figure were of this design. The bayonet was the most basic of all infantry weapons and yet, when compared to the machine gun and artillery which were the dominant forces on the battlefield, it would come to be relied on and prove itself invaluable in so many different ways. In 1915 the Russian army placed an order for 1.5 million Mosin–Nagant rifles and bayonets with the American armaments manufacturer Remington and an order for a further 1.8 million rifles and bayonets with Westinghouse. If one takes the 5 per cent figure of saw back bayonets for this order, then the Russian Army would have received 3.3 million bayonets of which 166,500 would have been the saw back type.

The effectiveness of the machine gun and artillery as great killing machines would be proven many times during the First World War. These weapons would come to be relied on to support an offensive such as the Third Battle of Ypres, also known as Passchendaele, between 3 July and 10 November 1917. During battle the artillery fired non–stop and at one point during a nineteen-day period the British guns 4.3 million shells of all calibre equating to some 107,000 tons. Despite this and other demonstrations as to how devastating these weapons were against infantry advancing across the open there were still commanders who held on tenaciously to the adage that the bayonet charge could carry the day. This was just as out-dated as having cavalry on the modern battlefield in the face of these weapons, also. Yet in the case of the cavalry it would also be wire, either barbed or plain, which would stop the horses, but commanders still hope for the so-called 'G in the Gap' when the cavalry could pour through and exploit the manoeuvre. It would never happen, but bayonet charges were mounted and troops fought bayonet tip to bayonet tip as their predecessors had in the Crimean War just sixty years earlier. This war also gave inspiration to poets caught up in the war and they mentioned bayonets in the lines of their verse. Popular songs of the time also mention the bayonet and even lesser-known songs and poems such as 'In the Morning', 'The Hipe', 'The Trench' and 'A Vision' make reference to it in some degree. For example in the Kiplingesque 'On Active Service' the lines include reference to the range of weapons including machine guns, poison gas, hand grenade and mortar, but it is the lines about the bayonet which are most poignant:

> 'W'en you're up against a feller with a bayonet long an' keen,
> Just 'ave purchase of your weapon an' you'll drill the beggar clean.
> W'en man and 'oss is chargin' you, upon your knees you kneel,
> An' catch the 'oss's breastbone with an inch or two of steel.
> It's sure to end its canter, an' as the creature stops
> The rider pitches forward an' you catch 'im as 'e drops.'

The bayonet drill being referred to was for use in a modern war but its techniques were from more than a century earlier and had been applied on the battlefield at Waterloo in 1815. The weapon to which the bayonet was

Long thrust upwards against trench raid.   Lunging at the throat.

German Army rifles
with bayonets used in
the First World War.

attached had changed and indeed so had the shape of the bayonet, but its method of use was unaltered.

During the First World War the three main belligerents in the west each had their own design in bayonets. The French Lebel rifle of 8mm calibre could be fitted with a bayonet to which the infantry gave the affectionate nickname of 'Rosalie' and was distinguished by its long, spike-like blade. The British .303 inch calibre Lee-Enfield rifle was equipped with a long knife-like bayonet, whilst the German army 7.92mm calibre Mauser rifle was fitted to accept a heavy-bladed bayonet which some thought looked more like a 'butcher's knife'. The length of bayonet blades would vary as the war progressed and the great lengths remained throughout, along with the propagandists' claims that the saw-back designs were being used to inflict horrendous wounds and the design went against all protocols laid out in the Geneva Convention. The simple fact remained that bayonets such as this were not outlawed by the Convention and secondly saw-backed bayonets were issued for use by troops serving with pioneer and engineer units and some sources claim that perhaps on 5 per cent of bayonets issued were of the saw back type. While proclaiming such contraventions of the Geneva Convention the propagandists very conveniently avoiding mentioning that these same saw-back type bayonets were also used by British troops and other Allied armies. Indeed, the Belgian

Army and even the Swiss Army, which was a neutral force, issued now back bayonets to pioneer units.

A contrast was made in 1915 which compared the British Lee-Enfield rifle and bayonet combination to the German K98 rifle with bayonet by stating that: 'The German soldier has eight inches the better of an argument over the British soldier when it comes to crossing bayonets, and the extra eight inches turns the battle in favour of the longer, if both men are of equal skill' This was correct because the British Lee-Enfield rifle measured 44.57 inches and the blade of the bayonet had a length of just over 20 inches to give and overall length of 61.57 inches. The German Modell 1888 Gewher (rifle) measured 48.80 inches in length and blade of its bayonet was 20.47 inches to give the combination an overall length of 69.27 inches making it around eight inches longer than its British rival. The French 'Rosalie' bayonet and its 20-inch cruciform section-blade fitted to the Lebel Modèle 1886 rifle measuring 51 inches produced a rifle and bayonet combination of some 71 inches (almost

Making ready to thrust upwards against trench attacker.

Ready to thrust upwards.

French infantryman with Lebel rifle circa 1914 with 'Rosalie' bayonet fitted.

French 'Poilou' with Lebel and 'Rosalie' bayonet in studio photograph.

British army rifles with bayonets used in the First World War.

six feet in length. The German army reduced the length of its bayonets and eventually stopped using the saw-back bayonet in 1917. Indeed, the saw back style was not very effective at cutting wood and pioneer battalions would have had access to proper saw for cutting wood. The shorter-bladed knife-type bayonets were better suited to trench warfare but some nations such as the French did not see any reason to change the style of their bayonets. In fact it would remain in use until the Second World War.

French soldier circa 19th century with 'Rosalie-type' bayonet, which remained in service until the Second World War.

As an offensive weapon the rifle and bayonet combination was still believed to be suitable for assaulting trenches, but using the great lengths in the confined space of a trench proved it was difficult to wield. Troops developed a range of hand-held weapons for trench raids and these included wooden clubs studded with nails, knives and daggers, and even short blades attached to 'knuckle dusters'. Such weapons were reminiscent of medieval warfare but the range of such weapons proved best suited for such close-quarter fighting

French troops and colonial troops, circa 19th century. They are equipped with the long-serving 'Rosalie' bayonet.

in confined spaces. Bayonets turned out to be better suited to the defence to lunge at attackers as they tried to jump into trenches. That applied to all sides, including the Americans when they entered the war in 1917. There were occasions when a man could use his bayonet in the dash across 'No Man's Land' when an individual enemy was encountered. A veteran of the First World War remembered using the bayonet and recalled how: 'I just saw him as an enemy that had to be defeated and at the time I sunk my bayonet into his body I didn't really give it too much thought. But when the time came to pull out my bayonet I found that it was quite difficult and so… I had to fire my rifle into his chest so that at the same time I could pull my bayonet out. I think I wasn't satisfied, I used the butt of my rifle and struck him somewhere on the head, I don't know exactly, to make sure it was all over.' It was just as Prince Frederick had seen when he observed soldiers using the butt and an example of how a man could revert to basic type using blades and clubs to kill an enemy.

In 1915 lance-corporal Stephen Westman was serving with an infantry regiment in the German army on the Western Front in France when he received the order to attack: 'We got the order to storm a French position, strongly held by the enemy and during the ensuing melèe a French corporal suddenly stood before me, both our bayonets at the ready, he to kill me, I to kill him. Sabre duels in Freiburg had taught me to be quicker than he and pushing his weapon aside I stabbed him through the chest. He dropped his rifle and fell, and the blood shot out of his mouth. I stood over him for a few seconds and then I gave him the *coup de grace*. After we had taken the position, I felt giddy, my knees shook, and I was actually sick' This was fighting up close and personal as it had been several centuries earlier in the Medieval period, but this was war in the

Short forward thrust for fighting in trenches.

Recreated scene showing how British infantry went 'over the top'.

twentieth century in an age when killing could be done at many miles range by heavy artillery. The late British actor Arnold Ridley, best known as his character of Private Godfrey in the popular television comedy series 'Dad's Army', fought on the Somme in 1915 and received two bayonet wounds. As a young soldier of 20 years of age he was serving in the Somerset Light Infantry and was engaged in attacking a German trench. A defender lunged at him and using his training he managed to deflect the thrust away from his stomach but the blade entered his groin instead. He had already received a heavy blow to the head, possibly by a rifle butt, and he passed out. As he was regaining consciousness he saw another German thrusting down at him with a bayonet. Ridley raised his hand and was bayoneted through the wrist. He was fortunate that he was rescued and evacuated before finally being medically discharged from the army in 1916 due to his wounds.

In 1917 in another theatre of fighting half a world away the Welsh-born author Ion Llewellyn Idriess was serving with the 5th Australian Light Horse at the first Battle of Gaza on 26 March 1917. The Light Horse regiments were mounted infantry, like modern Dragoons, who rode to the battle but fought on foot using rifles as infantry. At this battle they were part of a British force facing a Turko-German force of 4,000. Idriess recalled the fighting as: 'just berserk slaughter. A man sprang at the closest Turk and thrust and sprang aside and thrust again and again… the grunting breaths, the gritting teeth and the staring eyes of the Turk, the sobbing scream as the bayonet ripped home… Bayonet fighting is indescribable – a man's emotions race at feverish speed and afterwards words are incapable of describing feelings.' Westman had vomited after his face-to-face bayonet fight and after his encounter, Idriess became almost lost for words. It does not become more personal on a battlefield than stabbing with a bayonet and the experience affects different people in different ways. Combat affects different soldiers in different ways but something as

Thrusting upwards against an attacker.

personal as pushing a bayonet into someone is unlikely to ever be forgotten. But in the heat of battle when the 'red mist' descends a soldier goes into automatic mode and instinctively carries out the fighting drills taught to him in training. Australian troops in the First World War appeared to be more willing to use the bayonet than might be imagined as one man remembered on the Western Front in France when they: 'had order to bayonet all wounded Germans and they received it hot and strong.'

Several months later in November a number of mounted units of regiments in the British army were used to attack Turkish positions at El Mughar Ridge in Gaza. On 13 November six squadrons were ordered forward including the Dorset Yeomanry and the Buckinghamshire Yeomanry. They advanced up a steep rocky ridge for 200 yards all the time under constant heavy fire from machine guns. Two squadrons dismounted and fixed bayonets to attack the positions as infantry. They attacked and took the positions during the course of which they killed hundreds of Turkish troops captured many hundreds more and much equipment including machine guns. It was an extraordinary

US 90th Division full gear.

US 77th Division with bayonet and painted helmet 1917.

feat of arms with cavalry changing roles to serve as infantry and then mount a bayonet charge. It could have been a disaster but the shock impact must have been too great and the Turkish positions where overrun to neutralize the machine gun fire which would have killed many men and horse. When the war ended in November 1918 armies had learned a great deal and seen many changes such as the emergence of tanks, the use of poison gas and the dominance of aircraft over the battlefield to direct artillery fire. Yet, for all that, the bayonet attached to the rifle remained one of the strongest symbols of the war following more than four years of fighting.

Just twenty years later the world was heading towards another global conflict but in that intervening period there had been many wars in which the bayonet was still being wielded. Britain had troubles in India, France was engaged with its colonies in North Africa and the Far East and joining these nations was Italy under Benito Mussolini, who had been elected to power in 1922. In 1935 Italy invaded Abyssinia whose army had practiced bayonet fighting but ultimately proved no match against the Italians who used poison gas to kill

US Doughboy with backpack showing
bayonet being carried. It was difficult to
retrieve from this position.

US Doughboy of 77th Division in full gear.

British infantry practice bayonet fighting in training depot.

US Doughboys with helmets, gas masks and Enfields with bayonets, circa 1917.

US infantryman 'Doughboy' at bayonet practice, circa 1917.

soldiers and civilians indiscriminately. Mussolini boasted of Italy having eight million bayonets at its disposal but in the end it did not amount to much more than rhetoric. The Italian soldier had some good fighting qualities but poor supplies affected morale and fighting ability suffered as a consequence. Even so, Italian troops fought on in face of adversity. In 1936 Civil War broke out in Spain and Mussolini sent troops to support Nationalist General Franco and Hitler also sent German troops to help the cause. Italian troops did not prove effective in France in 1940 and against the Greeks they suffered badly. However, in Russia the Italian troops did fight well but in all cases there was only limited opportunity for using the bayonet. Japan had been flexing its military muscles since 1905 after defeating the Russians at Port Arthur and winning the Russo-Japanese War. The country had been allied to Britain in the First World War and seized much German territory in the Far East. In 1937 Japan invaded China and attacked great cities such as Nanking and Shanghai. The Chinese were numerically superior but the Japanese had tanks and aircraft to support them along with artillery. In the aftermath

Chinese soldier with rifle and fixed bayonet during the Sino–Japanese War in 1937.

of capturing these great metropolises the Japanese embarked in a series of atrocities during which they bayoneted, shot and decapitated hundreds of thousands of civilians and soldiers alike. The Japanese believed in Bushido (Way of the Warrior) as exemplified by the Samurai and the use of the sword in war. The bayonet as a bladed weapon should have followed this idea and been used in an honourable way in battle against armed enemies. Instead, in these cases and many other examples, the bayonet was not used as a weapon for the battlefield but in these circumstances it was being used in a murderous manner.

When the Second World War began in September 1939 the armies entering the field were armed with bayonets, but they also had at their disposal anti-tank guns, armoured divisions and fleets of aircraft to support them along with machine guns, artillery and mortars. Armies had never been mobilized with such firepower, which meant that during the war there would never be any requirement for full-on bayonet charges over open ground. There were occasions, when troops advanced under cover of an artillery barrage with bayonets fixed to deal with any small pockets of resistance they encountered, when butts and bayonets were used in the same manner as they had been centuries earlier. In his work *Soldier In Battle*, which appeared in 1941, Captain G.D. Mitchell. MC, DCM, suggested: 'The use of bayonet and butt go hand in hand. Lest any crevice be allowed for a sense of inferiority to creep in you must become expert in handling both--- just for emergency.' Basic training would have taught troops how to handle the bayonet and use the butt, but only in actual battle could such tactics be put to the test. It was still used in its traditional role to stab the enemy at close quarter, as a British army officer by the name of H.P. Samwell, who took part in the Battle of El Alamein, discovered very quickly. In his first-hand account of an incident which took place during the heat of the battle on 23 October 1942, he recalled how during an attack on a German position, some of them emerged with the apparent intention of surrendering. Instead one of the Germans threw a hand grenade. The German then took shelter in a trench and Samwell fired at him using his service pistol. When he ran out of ammunition he did not stop to reload but instead dropped the weapon and picked up a rifle which must have been fitted with a bayonet because he 'bayoneted two more and then came out again. I was quite cool now, and I started looking for my pistol, and thinking to myself there will be hell to pay with the quartermaster if I can't account for it.' It was an instinctive act to arm oneself with anything that came immediately to hand. In this case it was a rifle and he knew exactly how to use it with the bayonet. It is interesting that afterwards all he could concern himself with was the recovery of his pistol. Men do very strange things in the heat of battle and this was one such case.

The Austrian Army in the nineteenth century had used a tactic called 'stosstaktik' which was an attack in force. Such strategy looked powerful but in reality it was a costly affair, as shown at the Battle of Koniggratz in 1866. Powerful weapons inflicted huge casualties before such attacks had advanced very far across the battlefield as the British had experienced during the Battle of the Somme in July 1916. Indeed, by the time of the Second World War most armies had come to recognize the futility of such charges and replaced them with more considered tactics such as attacking from the flanks. The Russians did mount massed wave attacks against the Germans on the Eastern Front but it was the Japanese in the Far East where they used attacks called 'Tenno Heika Banzai' (Long live the Emperor) shortened to simply 'Banzai', which took these attacks to unprecedented levels. These infantry attacks with

Recreated German soldier with K98 rifle and his bayonet on his waist belt.

troops charging forward with fixed bayonets were suicidal and were intended to overwhelm opponents with the speed and ferocity. It was also hoped that such an attack would weaken morale by un-nerving the enemy. British and Americans discovered that if they held their ground and used all the weapons at their disposal they could defeat such a wild bayonet charge. Those Japanese who did get through to the Allied positions were met with bayonets and automatic fire at close quarters. Nevertheless, the Japanese continued to mount such fanatical attacks even against superior odds on island garrisons such as Iwo Jima and Tarawa.

The style of bayonets changed also during the war with the great lengths being reduced to knife-type weapons. The long-bladed styles were still kept by some armies and the British Home Guard kept their long bayonets for most of the war. The Germans reduced their bayonets and the British followed suit and even developed a 'spike-type' bayonet which was never popular and often referred to as the 'pig-sticker' and used on the No 4 Lee-Enfield rifle. The Americans developed shorter-bladed bayonets for their new range of weapons including the M1 Garand rifle. The bayonet was still being used in its traditional offensive and defensive roles, but it was now being used in a less aggressive role to guard prisoners of war. The presence of it fitted to a rifle seemed sufficient to impress on the surrendered troops that their guards were in control. The bayonet had long been used as a multi-purpose tool and in the Second World War it was to put to use as a mine detector, especially in North Africa where millions of mines were laid by both sides. Because so many mines were laid over miles the engineers who were given the task of clearing paths through the minefields for the tanks to pass through did not have enough specialist electronic detectors to complete the job. The infantry were given instruction on how to use their bayonets held at an angle to probe the sand very lightly to detect the mines. Once a device was located its position was marked and engineers made it safe. Hundreds of troops used in this way made it possible to clear paths through minefields. The British infantry were equipped with wooden-handles for their entrenching tools but it was also possible to fit the spike-type bayonet to the end for use in probing for mines. The tactic proved successful and every army would use it eventually for this purpose. The Germans used this tactic on the Russian Front to remove mines before attacking Soviet positions. In fact, the method is still used today and bayonets remain an invaluable tool for this purpose in places such as Iraq and Afghanistan.

On the 6 August 1945 the USAAF dropped the first atomic bomb to be used in war on the Japanese city of Hiroshima. Three days later another atomic bomb was dropped, this time on the Japanese city of Nagasaki and these two events led to the end of the Second World War. The world had now entered a new era and the face of warfare was changed forever. Atomic weapons were reduced in size and capable of being fitted to short-range battlefield rockets to provide armies with tactical weapons. This gave armies a new weapon and created a trinity of weapons of mass destruction along with chemical and biological weapons. These last two types had been used for centuries in crude forms but by the Second World War they had been refined to the point where they could threaten entire cities. Together these three types of weapon were called 'nuclear, biological and chemical' or NBC for short. Troops had to be issued with special protective clothing and tactics had to be developed which would allow troops to continue military operations in the event of an enemy using NBC weapons. This was Armageddon or the ultimate form of destruction. The use of a single atomic weapon can kill thousands of people in a massive blast and chemical and biological weapons have the potential to cause long-term mortality. Yet for all this destructive power the infantry still carry bayonets. Countries such as America, China, Britain and France along with several other countries have nuclear weapons, but they still issue bayonets to their troops. These are the same soldiers who can fire missiles to destroy a tank or an aircraft using high-technology precision and then be expected to fix their bayonets and get in close with the enemy. This scenario happened during the Falklands War in 1982 when British troops were using missiles such as the shoulder-fired Blowpipe to shoot down aircraft and Milan anti-tank missiles to clear strongpoints. Supersonic aircraft were launching missiles to destroy ships and amidst all this the men on the ground still fixed bayonets to get in close with the Argentine troops. Science has created the most effective sophisticated weaponry and yet soldiers still carry with them a weapon which has not changed that much in over 350 years. Rifles and other types of firearms in service today are still used in the same way as the much older types of weapons they have replaced, which have long since been taken out of service. The bayonet is another example of this rule and is still used to stab the enemy to death in the same way soldiers did using socket bayonets in battles such as Culloden, Blenheim, Waterloo, Inkerman and the thousands of other battles fought over the centuries.

In the many post-war conflicts around the world troops have still found it necessary to use the bayonet on occasion. The Korean War began in 1950 and continued until 1953 when the North Korean army attacked South Korean. The forces of the United Nations, including Australia, America, Britain, Canada and France all sent troops to the defence of South Korea. The North Koreans were supported by Communist China and much of the equipment used by all armies dated from the Second World War. Bayonet charges and bayonet fighting took place in some actions such as the attack led by Captain (later Colonel) Lewis Millett of the US 27th infantry Regiment. On 7 February 1951 at Hill 180 near Soam-Ni a platoon of US infantrymen became pinned down by heavy enemy fire coming from the feature of Hill 180. Captain Millett was an experienced soldier, having seen service in Italy during the Second World War, moved his platoon forward, collecting more men as he did so, and with this assembly he charged up the hill. It was a desperate action, throwing hand grenades as they charged. After the fight, 50 enemy dead were counted, of which 20 were found to have been killed by bayonet thrusts. This equates to 40 per cent casualties, which is a much higher figure than average. That does not mean the bayonet was used more often during the Korean War, but was just what happened in this one particular incident. The historian S.L.A. Marshall described the action as: 'the most

French troops with Mistral anti-aicraft missile carrying bayonets for their FAMAS rifles.

complete bayonet charge by American troops since Cold Harbor'. This was a comparison to the battle in June 1864 during the American Civil War. Marshall may have overpraised the fact because it was the only true bayonet charge of his generation. That does not diminish the actions of Millett who was awarded the Medal of Honor for his action.

The role of combat service for the bayonet continues in modern times such as action by French troops of the 3rd Regiment d'Infanterie de Marine in 1995. This unit was serving in Sarajevo when they fixed bayonets to charge Serbian-held positions of the Vrbanja Bridge to clear them away and secure their own positions. These actions are more bayonet fights rather than actual bayonet charges in the historical interpretation of the term. Nevertheless, they still illustrate how important the bayonet still is today and the story goes on. In 2009, Lieutenant James Adamson of the Royal Regiment of Scotland was awarded the Military Cross for his part in a bayonet fight in Afghanistan. Adamson had already shot and killed one Taliban fighter when a second man appeared, whom he bayoneted. More recently, Lance Corporal Sean Jones of the Princess of Wales's Royal Regiment also received the Military Cross

Accessories for Swiss SG542 including hollow–handle bayonet.

Danish troops leaving their APC, who are equipped with bayonets.

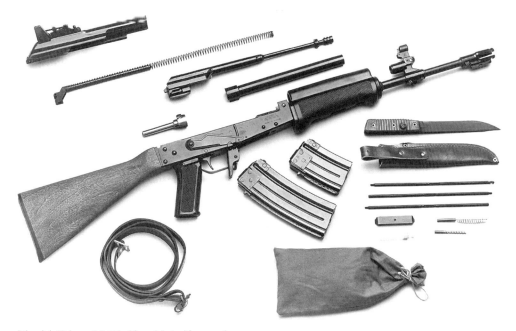

Finnish Valmet M-76 rifle with knife-type bayonet.

in 2012 for his part in a bayonet fight in Afghanistan in October 2011. Some armies have bayonets permanently fixed as part of their standard operating procedure for troops on patrol when deployed to operational areas such as Afghanistan. No doubt this will remain the practice for future deployments due to the usefulness of the bayonet in such situations.

Modern bayonets are produced to an exacting high standard using the best materials in order to ensure reliability under all conditions and extremes of temperatures from sub-zero Arctic conditions to high humidity and the arid desert wastes. Some parts that were traditionally produced using wood, such as the grips on the handle are now produced using very advanced plastics, which can withstand the rigours of the modern battlefield. Designers and producers have also taken notice of how bayonets have traditionally been used as a tool for all manner of tasks. Bayonets are made in a range of styles that includes types which fit flush with the barrel, those with a barrel ring in the quillon and the third type which is the 'tubular' hilt and is almost like a reverting back to the socket bayonet. The British Army uses a similar style hollow-handled bayonet on the SA-80 rifle, which fits over the barrel and wraps around the muzzle. These are unusual styles but regardless of how the bayonet looks the primary function of it as a weapon for close-quarter fighting still remains.

Perhaps because of the simplicity of its design the bayonet has been allowed to remain in service for so long. It seems remarkable that after all these centuries the bayonet is still available to the infantry as a practical and functional item as well as being extremely versatile in a modern age and continues to serve as a multi-purpose tool and a cooking implement. This usefulness in battle is enhanced by night vision goggles, which allow soldiers to see clearly even on the darkest nights, to enable fighting 24 hours a day in all weather conditions and permits the use of the bayonet in such conditions.

Norwegian soldier with night vision goggles and carrying a bayonet.

Designers have reverted to type and modern bayonets have lengths that rarely exceed 10 to 12 inches, and the style is of the knife-type with a drop point. Manufacturers put their bayonets through a series of rigorous tests to check they meet exacting standards and will perform under all conditions. For example, blades a subjected to temperatures of -40 degrees C and then used to lever open boxes to test if the blade shatters under stress. Keeping the bayonet at sub-zero temperatures it is fitted to a rifle and which is fired to check what effect sudden blasts of heat during the firing process has on the deep-frozen blades. Under normal ambient temperatures bayonet blades will remain unaffected by more than 5,000 rounds being fired at a cyclic rate of up to 900 rounds per minute. This not only tests the durability of the blade but the locking mechanism, also. Troops operate in extreme heat and blades are tested to +80 degrees C

Freeze-testing modern bayonets and then firing.

and bayonets are put through a range of stress and firing tests to check the integrity of the blade.

It has been discovered that a range of substances can affect a bayonet and designs are subjected to salt water, corrosive chemical tests and fuel oils. All bayonets are today of a type which automatically lock onto the rifle without the need to depress a spring catch. However, in order to remove it the catch has to be positively depressed to release the bayonet. It is this spring-operated part which can cause problems and is why stringent tests are routinely completed to prevent failure. Even the simple task of removing the bayonet from the scabbard is tested to make sure it does not stick and also measure the amount of energy required to withdraw it. If the bayonet is too loose it may become lost as a soldier runs around, and if it is too tight a soldier could experience problems of drawing it from the scabbard. A fine line has to be achieved for it to meet operational levels that are acceptable for modern battle conditions.

The modern bayonet design is for the most part knife-type but some blades are manufactured with spear points like the Fairbairn-Sykes Commando knives of the Second World War and these can be fitted to a range of modern rifles such as the M16 and G3. Even the modern 'Bullpup' styles where the

Norwegian army winter warfare still carry bayonets.

magazine is mounted behind the pistol grip, such as the French FAMAs, British SA-80, Chinese QBZ-95 and the Austrian AUG, the shape of the barrels of these rifles allows for the fitting of bayonets. Blades remain traditional in style, being flat usually with a single cutting edge. Some types have double edges and the scabbards have a small area of special surface attached for re-sharpening the blades. Some designs of blades have a fine-toothed saw edge along the back to allow sheets of toughened plastic or metal sheeting to be cut in an emergency. This is similar to the old 'saw-backed' pioneer bayonets, but the modern styles are more subtle and not so immediately obvious. The blades of some modern bayonets have a small piece cut out at some point along the length and this interlocks with a lug on the scabbard, which allows the bayonet to be converted into a useful wire-cutter. Scabbards are for the most part made from high-impact heavy-duty plastics, although some are still manufactured in metal for those armies which express a preference for this type. The chape of some types of scabbard are fitted with a small metal attachment which is fixed in place with small screws, and this often terminates in a chisel point which can be used to undo slot-headed screws. It is at this

US soldier about to fire an M47 'Dragon' anti-tank missile. He carries a bayonet for his rifle on his belt.

Modern bayonets can be used as wire cutters.

point the lug for connecting to the cut away on the bayonet blade to turn it into a wire cutter is located. Some manufacturers have also seen fit to include a bottle-opener in their designs in an ironic move from the days when troops used their bayonets to open bottles.

Soldiers have a tradition of using whatever means they have at hand in order to achieve something and that includes bayonets which they have always used as a handy tool especially to lever open wooden cases and chop wood. The shape of the modern bayonet also lends itself to being used as a hammer and even breaking glass in windows to allow soldiers to enter a building when fighting in built-up areas. In an emergency soldiers today can still use their bayonets to probe for mines just as the troops did in the Second World War. The scabbards of modern bayonets have been developed to allow them to interconnect with the blade of the bayonet and be used as a wire cutter,

Probing for mines with bayonets is still being practised.

Modern bayonets are still used as basic tools for opening crates.

extending its use as a specialized tool. This allows a soldier to cut through barbed wire and cables to remove obstacles. Indeed, some manufacturers have recognized the fact that wires may be electrified thereby presenting a danger to soldiers and they have fitted a small indicator to the scabbard to show if a wire has an electric current running through it. This is like an electrician's device and can indicate voltages from 70 to 400 volts. The plastic scabbards and plastic handles on the bayonets act as insulators but this small device is an added precaution. A leading German bayonet manufacturer

Weapon systems from FN Herstal, Belgium, showing solid- and hollow-handle bayonets for rifles.

Swiss-made SG 542 rifle, which can be fitted with bayonet.

Swiss-made SG5-42 rifle with hollow-handle bayonet fitted and bipod support.

summed it up when they stated: 'a good bayonet is an invaluable asset without increasing the weight of a soldier's load compared with a combat knife.' In other words, a knife is a knife but a bayonet is a multifunctional tool.

The bayonet has been in service now for over 350 years during which time it has become a reliable item in the infantryman's personal weaponry. It has changed dramatically in that time but it has always been there to be put to use in a variety of roles including that of hand tool for cutting or to hammer objects. The bayonet itself is still manufactured from steel although some modern alloys are now used which are just as strong but much lighter weight. Traditional wooden parts such as the handle grips have been replaced by plastic, which is more durable and is not affected by extremes of climate. Some research has been undertaken to try and develop an even lighter bayonet, not that modern bayonets are particularly heavy. These trials have investigated the use of tougher plastics and other alloys. One idea is that because most of the impact of a thrust is taken by the tip of the blade only this portion of the bayonet needs to be metal. This point can be made from tungsten, which is extremely hard metal, and the rest can be made from plastic for ease of production. Unfortunately it is not that straightforward because soldiers will still try to use the bayonet to lever open cases and the pressure exerted would snap the blade. Another concept is a return to the 'pig sticker' spike-type bayonet, but even this would not work. History shows that such a design was never popular and few such types ever entered full service. Such a design would be prone to bending and if made from plastic tipped with tungsten it would no doubt snap, even if made using high grade plastic. The bayonet is one of those rare weapons the design of which does not have to be unduly altered or modified. Full plastic bayonets are about as unlikely as rifles made completely from plastic. Small component parts made from plastic can be used but overall the design has to be metal for durability and strength. It is

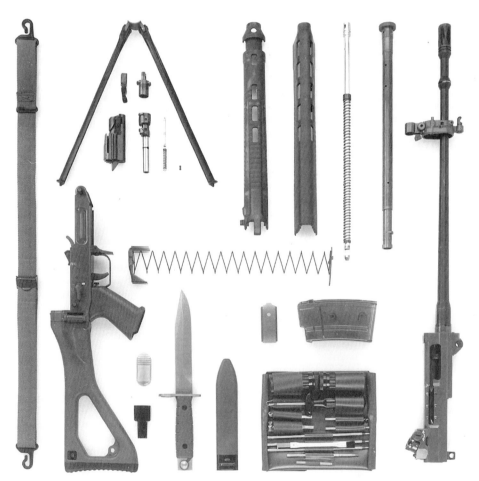

The Swiss-made SG 550 rifle showing knife-type bayonet with solid handle.

unlikely that more radical changes will be made to the design of the bayonet, and even the shape of modern designs all follow a similar shape, which is usually knife-type with drop point. As long as infantry forms part of armies there will always be a need for the bayonet and it will always be used as shown during the Falklands War of 1982 and more recently in Iraq and Afghanistan.

As the use of the bayonet continues well into the twenty-first century, anyone curious about how much longer it will survive in this day and age of technological weaponry would do well to think about the words of Edward Costello, who fought as a rifleman in the 95th Rifles during the Peninsular War in 1812, and who wrote rather prophetically that: 'the bayonets had better remain in present use until such time as we can bargain with the French or other enemies to disuse them.' Two hundred years later, these words have a strange resonance to them as troops of the United Nations serve in Afghanistan where the bayonet is still being used for the purpose for which it was intended. In conclusion, according to the nineteenth century historian Prince Kraft zu Hohenlohe-Ingelfingen: 'he who has not made up his mind to come at last to the bayonet can never win, for he can have no serious intention to assault.' With that thought in mind, we can be certain that the bayonet will continue to be used for many more years to come, no matter what advances are made in weaponry design.

Where to see Bayonets

Many military museums around the world have bayonets on display, including the various regimental and national collections. These are all worth visiting to discover bayonets and view the various designs. This list is an introductory guide to some of the more important collections.

The National Army Museum, Chelsea, London. www.nam.ac.uk

Royal Armouries, Leeds. www.royalarmouries.org

Various regimental museums around the country. A list is available through the Ogilby Trust at: www.armymuseums.org.uk

Cobbaton Combat Collection, Chittlehampton, Umberleigh, North Devon EX37 9RZ. Telephone 01769 540740. www.cobbatoncombat.co.uk

The Wallace Collection, Hertford House, Manchester Square, London W1U 3BN. Telephone 0207 563 9500. www.wallacecollection.org

Hotel National de Invalides Army Museum in Paris, France: www.musee-armee.fr

Some Useful Bayonet Websites

History of bayonets, identification of bayonets and illustrations and features: www.worldbayonet.com

Interesting features and illustrations: www.holmback.se/bayonet/links.html

Website of the Society of American Bayonet Collectors: www.bayonetcollectors.org

Buying and collecting bayonets of the world: www.bayonetconnection.com

Infantry tactics, weapons and equipment of the armies of the Napoleonic Wars: www.napolun.com

Bibliography

Barnes. Major R.M: Uniform and History of the Scottish Regiments, Sphere Books Ltd, London 1972.

Bull. Stephen: An Historical Guide to Arms and Armour. Studio Editions Ltd, London 1991.

Casson. Stanley: Steady Drummer. G. Bell & Sons, London 1935.

Carver, Field Marshal Lord: The Seven Ages of the British Army. Grafton Books, London 1986.

Cleator. P.E: Weapons of War. Robert Hale, London 1967.

Cross. Robin (General Editor): The Guinness Encyclopedia of Warfare, Guinness Publishing Ltd, London, 1991.

David. Saul: All the King's Men. Viking, Penguin, London, 2012.

Davis. William C: The Illustrated Encyclopaedia of the Civil War, Salamander Books Ltd, London 2001.

Dempsey. Guy: Albuera 1811; The Bloodiest Battle of the Peninsular War. Frontline Books, Barnsley, South Yorkshire, 2011.

Dyer. Gwynne: War, Guild Publishing, London 1986.

Evans. Roger D.C. and Stephens, Frederick J.: The Bayonet: An Evolution and History, Milton Keynes 1985.

Ffoulkes. Charles: Arms & Armament, George G. Harrap and Company Ltd, London, 1945.

Fuller. J.F.C: Decisive Battles of the Western World, Granada Publishing Ltd, London, 1970.

Haswell. Jock: The British Army; A Concise History. Thames & Hudson, London 1975.

Hogg. Ian V.: The Encyclopedia of Weaponry, Greenwich Editions, London 1998.

Hogg. Ian.V and Weeks. John: Military Small Arms of the 20th Century, Arms and armour Press, London, 1977.

Holmes. Richard: Firing Line, Jonathan Cape Ltd, London 1985.

Holmes. Richard: Sahib; The British Soldier in India, Harper Collins, London 2005.

Holmes. Richard (et al): The World Atlas of Warfare, Mitchell Beazley, London, 1988.

Holmes. Richard: Redcoat, Harper Collins, London 2001.

Isemonger. Paul Lewis and Scott. Christopher: The Fighting Man. Sutton Publishing, Stroud, Gloucestershire 1998.

Keegan. John and Holmes. Richard: Soldiers; A History of Men in Battle, Sphere Books Ltd, London, 1987.

Kerr. Paul (et al): The Crimean War, Boxtree, London 1997.

Laffin. John: A Western Front Companion 1914-1918, Sutton Publishing, Stroud, Gloucestershire, 1994.

Longmate. Norman: Island Fortress; The Defence of Great Britain 1603-1945, Grafton, London 1993.

Massie. Alastair: The Crimean War; The Untold Stories, Sidgwick & Jackson, London, 2004.

Montgomery of Alamein: The Memoirs, Collins, London, 1958.

Montgomery of Alamein: A History of Warfare, Collins, London 1968.

Newark. Peter: Firefight! The History of Personal Firepower, David & Charles, Newton Abbot, 1989.

Norris. John: Artillery; A History, Sutton Publishing Ltd, Gloucestershire, 2000.

Peterson. Harold L: The Book of the Gun. Hamlyn, London 1963.

Pope. Dudley: Guns. The Hamlyn Publishing Group Ltd, London 1969.

Reid. William: The Lore of Arms. Purnell Book services Ltd, Oxon 1976.

Roger. Colonel H.C.B.: Weapons of the British Soldier, Seeley, Service & Co Ltd, London 1960.

Rothenberg. Gunther.E: The Art of Warfare in the Age of Napoleon, Spellmount, Kent 1997.

Townsend. Charles. Et al: The Oxford Illustrated History of Modern War.' Oxford University Press, Oxford, 1997.

Wilkinson. Frederick: The World's Great Guns, Hamlyn, London 1977.

Wilkinson Latham. R.J: British Military Bayonets from 1700 to 1945, Hutchinson & Co, London 1972.

Winter. Dennis: Death's Men; Soldiers of the Great War, Penguin Books, London, 1979.

Index

Act of Union (United Kingdom) 1707, 33

Adamson, Lt James (MC), xi, 193

Afghanistan, xi, 143, 191, 193, 195, 203

American Civil War, viii, 74, 89, 111, 115, 118–23, 138, 153, 170, 193

Argyll & Sutherland Highlanders, x

Austerlitz, Battle, 95

Austro-Prussian War, 103, 110, 124

Baker Rifle, 59–61, 112

Bayonet Scandal, 134–5

Bayonne, 4–7, 9–10

Bell, Sir Charles, 115, 174

Blenheim, Battle, 49, 192

Boyn, Battle, 87

Brandywine Creek, Battle, 81–2

Brown Bess musket, 52–3, 60–1, 66, 80, 105

Bunker Hill, Battle, 49, 78, 80

Burgoyne, Gen John, 62, 75–6

Buxar, Battle, 53

Canada, 37, 57, 71–2, 130, 192

Camerone, Battle, 1–3

Carnot, Lazare, 91

Charles I, King, 13–14

Charles II, King, 14–17, 21–2

Charleville musket, 53–4, 98, 105

Chassepot rifle, 104, 112, 125

Chastenet, Chevalier Jacques de, 10, 33, 37

Churchill, John (Duke of Marlborough), 22, 24, 46, 62

Clive, Col Robert, 62

Coehoorn, Baron Menno van, 37

Colt, Samuel, 103

Cope, Sir John, 67

Cranking the blade, 41, 49

Crimean War, viii, xi, 1–2, 51, 74, 89, 109–11, 113–15, 151, 176

Cromwell, Oliver, 13–14

Cuba, 74, 124, 135

Culloden, Battle, 37, 65–6, 69–70, 73, 192

Cumberland, Duke of, 37, 46, 65–6, 68–9

Cutlass bayonet, 133

Damascus steel, 5

Danjou, Capt Jean, 1–3

Dettingen, Battle, 11–12, 65

Donauworth, Battle, 62

Dragomirov, Gen Mikhail, 83

Dragoons, 8, 16, 18, 22, 76, 81, 183

Durs Egg, 42–3, 44, 56

Elcho bayonet, 107–108

Epée bayonet, 112, 134

Ewart, Sgt Charles, 100–101

Fairbairn-Sykes dagger, 7, 197

Falkirk, Battle, 65, 68, 87

Falklands, vii, x, 105, 143, 164, 192, 203

Ferguson, Patrick, 55–6

Flamberge style, 27, 106

Folding bayonets, 42, 44–5, 160, 165–6, 168–9

Franco-Prussian War, 89, 103–104, 108, 110, 124, 170

Fraustadt, Battle, 83–4

Frederick the Great, 49–50, 85

Fusiliers, 8, 14, 17, 19–20, 22, 24, 39, 71, 76, 107, 117

Grey, Maj-Gen Charles 'No Flint', 81
Guards Depot Pirbright, 137
Gurkhas, 105–107
Guthrie, George, 115, 148–9

Halberd bayonet, 22, 28–9, 33, 106–107
Henry Rifle, 108, 123, 125–31, 133–4

Indian Mutiny, 105, 111, 115–18, 123
Inkerman, Battle, x, 51, 113–15, 146, 151, 192
Iraq, x–xi, 143, 158, 191, 203

Jacobite Rebellion, 25, 65, 67–70, 72
James II, King, 21–5
Jersey, Channel Islands, 57
Jones, Cpl Sean MC, xi, 193

Karl XII, King of Sweden, 83–4
Katyn, 113, 168
Killiecrankie, Battle, 25–6
Kipling, Rudyard, 176–7
Korean War, 192
Kukri bayonet, 106–107
Kulm, Battle, 94
Kyhl, Johan C.W., 58

Laffin, John, 141–2
Larrey, Baron Dominique Jean, 115, 147, 174
Louis XIV, King of France, 15, 19, 32, 34, 36–8, 45–6, 91
Lunger blade, 121, 125, 129, 134

MacKay, Gen Hugh, 25–6
Mahdist War, 90, 102
Malplaquet, Battle, 63–4, 115
Marlborough, Duke of *see* Churchill, John
Marlborough's Wars, 12, 62
Marsaglia, Battle, 45–6

Martinet, Lt-Col Jean, 9–10
Martini-Henry rifle, 104, 108, 125–131, 133–4
Mexico, 1–3, 109, 118, 122
Miani, Battle, 58
Monmouth, Duke of, vi, 15, 21–2, 24, 62
Montgomery, Field Marshal Sir Bernard Law, 138
Morgan. Maj-Gen Thomas, 11
Moro Rebels, 131
Mussolini, Benito, 185–7

Napoleon Bonaparte, viii, 30, 47, 83, 92–8, 101–102, 108, 137, 144
Napoleon III, 1, 108–109
Naseby, Battle, 13
Needlegun, 103, 125
New Model Army, 13–14
Nine Years' War, 38, 44–5
Nock, Henry (Bayonet), 57

Oblique Order tactic, 50

Paraguayan War, 89
Peninsular War, 93, 148–9, 151, 203
Petit, Jean Louis, 145
Philippines, 131
Picq, Ardant du, 110
Pike, 8, 10–11, 13, 19–20, 28–33, 36–8, 46, 53, 73, 75, 86–7, 96, 106
Plains of Abraham, 72
Plug bayonet, ix, 4, 7–12, 15–16, 18, 20–1, 24, 26–8, 30, 33–7, 39, 44–6, 81
Port Arthur, 135 , 187
Prussian Army, 48–51, 78, 86, 97–8, 103, 110, 124, 137
Pyramids, Battle, 92

Quebec, Battle, 72

Ramrod bayonet, 136
Ridley, Arnold, 183
Romans, vii–viii, 5, 86, 133

Roosevelt, Theodore, 136
Rossbach, Battle, 88
Russo–Swedish War, 83

Schiess, Cpl Christian Ferdinand, 129
Schiltron tactic, 86–7
Scots Guards, vii, 105, 143, 167
Sedgemoor, Battle, 22
Solferino, Battle, 109, 157
Solingen, 5
Spanish Succession, war, 22–4, 32–3, 40, 62
Spear bayonet, 43–4
Stony Point, Battle, 79, 81
Suvorov, Gen Alexander, 83–5
Sword bayonet, 59–60, 104, 108, 112, 132–5

Talavera, Battle, 93–4
Tel El-Kebir, Battle, 128, 130
Tercio formation, 87
Thin Red Line, viii, 51, 114
Tiger bayonet, 86
Tipu Sultan, 86, 92
Trooping the Colour, 157–8
Trowel bayonet, 123

Ulundi, Battle, 89 90

Vauban, Marshal Sébastien Le Prestre de, 4, 32, 37–8, 40, 48
Von Steuben, Gen Friedrich, 78–9

Washington, Gen George, 75–6, 78, 80, 82
Waterloo, Battle, 60, 88–9, 98, 100–101, 113, 119, 130, 177, 192
Wellesley, Col Arthur (later Duke of Wellington), 92–4
William of Orange, Prince, 24–5
William III, King, 24–5, 47, 87
Wolfe, James, 46, 62, 64–5, 70–3
Wounding by bayonets, viii, x, 115, 118, 136, 144

Yataghan blade, 104, 112, 119, 121, 124–5

Zouaves, 124
Zulu War, 89, 102, 124, 126–7, 129–30, 134, 159
Zulu warriors, 89, 102, 126–7, 129